Advances in Fossil Energy R&D 2016

Topical Conference at the 2016 AIChE Annual Meeting

San Francisco, California, USA
13 - 18 November 2016

ISBN: 978-1-5108-3427-9

Printed from e-media with permission by:

Curran Associates, Inc.
57 Morehouse Lane
Red Hook, NY 12571

Some format issues inherent in the e-media version may also appear in this print version.

Printed by Curran Associates, Inc. (2017)

For permission requests, please contact AIChE
at the address below.

AIChE
120 Wall Street, FL 23
New York, NY 10005-4020

Phone: (800) 242-4363
Fax: (203) 775-5177

www.aiche.org

Additional copies of this publication are available from:

Curran Associates, Inc.
57 Morehouse Lane
Red Hook, NY 12571 USA
Phone: 845-758-0400
Fax: 845-758-2633
Email: curran@proceedings.com
Web: www.proceedings.com

TABLE OF CONTENTS

469126 Physical Solvent Mixtures for CO_2 Absorption at Simulated Pre-Combustion Conditions

Abstract:

We have developed physical solvent mixtures for CO_2 absorption at simulated pre-combustion conditions. In this study, the gas solubility and mass transfer coefficient of CO_2, H_2, CH_4 in a physical solvent, an ionic liquid, and other mixtures were measured in a Continuous Stirred Tank Reactor (CSTR) by the step-pressure dynamic method. The solvent performance and their mass-transfer properties in a gas-liquid system are important for engineering design and unit operation. The absorption of CO_2, and H_2 in physical solvents and other mixture were investigated. Effects of pressure on the volumetric coefficient K_La were studied with various gas liquid systems in the pressure range 0-30 atm. Experiment data of the mass transfer coefficient of CO_2 in physical solvents and mixtures matched predictions. Based on the experiment observation, physical properties and the kinetics of mixed solvents followed ideal solution mixing rules. Results of this study provided a base for further investigation on solvent mixture for gas absorption.

460582 High Performance, Bio-Derived Solvents for Pre-Combustion Capture

Abstract:

While Selexol/DMPEG is a well-established process/solvent for CO2 removal in applications such as natural gas processing and syngas purification, and is expected to be a leading candidate for pre-combustion CO2 capture applications, the viscosity of DMPEG is several times that of water. As such, DMPEG may face challenges when used within hollow fiber membrane contactors (HFMCs) and other advanced devices designed to promote high mass transfer rates. Recently, we have developed new types of physical solvents from bio-derived sources that exhibit much lower viscosities than DMPEG, low volatility and have a greater absorption capacity for CO2 at given temperature and pressure conditions. As such, this class of solvents appears highly promising for further scale up and demonstration projects. This presentation will discuss solvent production, physical property characterizations, molecular simulation and process modeling.

471696 New Water-Lean Amine-Based Solvents for Carbon Dioxide Capture with Low Regeneration Energy

Abstract:

Carbon dioxide capture and storage (CCS) of anthropogenic CO_2 released from large point sources such as coal fired power plants is required in order to mitigate adverse environmental effects like climate change. Toward this goal numerous CO_2 capture processes have been developed, the most advanced technology utilizes aqueous solutions of primary and secondary alkanolamines such as MEA or diethanolamine (DEA) for chemical absorption of CO_2. The regeneration energy of these aqueous amine solvents is very high due to the high heat capacity of water. To address this problem water-lean transformational solvent systems have been developed. As part of an ongoing program studying multiple solvent chemistries and combining the key findings from different solvent classes, we have developed a new class of water-lean non-aqueous amine-based solvents with potential to be energy efficient CO_2 capture materials. These amines bind CO_2 at standard temperature and pressure (STP) to form liquid carbamate salts with high gravimetric capacity ~ 20 wt%.Â These solvents have low thermal regeneration temperature (70-100 Â°C) and low CO_2â€“rich solvent viscosity compared to other water-lean materials. The regeneration temperature can be lowered to 60 Â°C by applying polarity swing assisted regeneration (PSAR)Â making it feasible to utilize lower grade heat from the power plant for CO_2 stripping resulting in energy savings.

461574 Molecular Design Strategies to Reduce the Viscosity of Non-Aqueous Carbon Capture Solvents

Abstract:

In power generation, high viscosities encumber the use of post-combustion non-aqueous CO_2 capture solvents. Using single molecule CO_2 binding organic liquids (CO_2BOLs) as example, the key molecular features that determine viscosity and CO_2 uptake kinetics are identified. The close proximity of the alcohol and amine sites involved in CO_2 binding result in the concerted formation of a Zwitterion containing both an alkyl-carbonate and a protonated amine. Capture is controlled by viscosity since the small binding energy barrier suggests that diffusion is rate limiting. The viscosity of CO_2BOLs increases exponentially as a function of CO_2 loading. Inter-molecular hydrogen bonding between positively and negatively charged functional groups ultimately determines viscosity. The hydrogen-bonding network can be controlled by chemically tuning the solvents to favor an internal hydrogen bonding structure. Additionally, a molecular design strategy to significantly reduce viscosity by shifting the proton transfer equilibrium toward a non-charged acid/amine species, as opposed to the ubiquitously accepted zwitterionic state, is presented. Based on knowledge obtained from molecular simulation, a reduced model was developed that predicts CO_2BOL viscosity given a few structural parameters and CO_2 loading. In collaboration with experiment, candidate compounds were synthesized to measure their viscosities and verify the model.

467008 Molecular Simulation of CO_2 Absorption in the Ionic Liquid [P2228][2-Cnpyr] Using Reaction Ensemble Monte Carlo

Abstract:

CO_2 capture using monoethanolamine (MEA) suffers from amine degradation, solvent emissions, and large heat requirements to regenerate the solvent. Ionic liquids (ILs) are an attractive alternative to MEA due to their thermal stability, minimal vapor pressure, and tunable properties. Gurkan et al. [1] designed a pyrrolide-based anion that binds CO_2 more weakly than MEA, but still absorbs CO_2 in near equimolar amounts at standard conditions. Further IL optimization is complicated by the exceptionally large number of possible cation-anion pairs. Molecular simulation could be a valuable tool to screen proposed ILs for CO_2 absorption, if it were possible to simulate both the condensed phase IL environment and the C-N bond formation as CO_2 binds to the pyrrolide anion. Quantum mechanical methods can accurately model bond formation and breaking, but are limited to small system sizes representative of the gas phase environment. Classical simulations, on the other hand, can be used to model the bulk liquid at a molecular level, but cannot describe covalent bond formation.Â Weâ€™ve developed a methodology for predicting CO_2 absorption in ILs using reaction ensemble Monte Carlo (RxMC) that utilizes both quantum and classical simulations. Liquid phase configurations are sampled using traditional translation, rotation and regrowth Monte Carlo moves. The equilibrium composition is sampled using reaction moves that delete reactant molecules and insert product molecules. The acceptance probability of reaction moves depends on the liquid-phase change in potential energy due to molecular insertions and deletions and the gas-phase change in free energy dueÂ to the reaction. The latter is determined by computing the potential of mean force using ab initio molecular dynamics. We investigate the absorption of CO_2 in the ionic liquid triethyloctylphosphonium 2-cyanopyrrolide, [P2228][2-CNpyr], and present CO_2 absorption isotherms at a range of temperatures.

[1] Gurkan, B., Goodrich, B.F., Mindrup, E.M., Ficke, L.E., Massel, M., Seo, S., Senftle, T.P., Wu, H., Glaser, M.F., Shah, J.K., Maginn, E.J., Brennecke J.F. and Schneider W.F., "Molecular design of high capacity, low viscosity,

chemically tunable ionic liquids for CO_2 capture." The Journal of Physical Chemistry Letters 1.24 (2010): 3494-3499.

472611 A Novel CO_2 Absorption Process Enabled By Dedicated Biphasic Solvents: Solvent Screening and Process Development

Abstract:

Biphasic solvent-based absorption processes have recently been of growing research interest as an alternative technology to the conventional absorption processes for post-combustion CO_2 capture. Such processes employ biphasic solvents that can form dual liquid phases during CO_2 absorption, with the absorbed CO_2 concentrated in one phase and lean in the other. The CO_2-concentrated liquid phase is the only part of solvent to be regenerated, resulting in a reduced mass for CO_2 stripping, an elevated stripping pressure associated with high CO_2 loading, and thus reduced energy consumption.

We are presently developing a novel absorption process enabled by a dedicated new class of biphasic solvents to further improve the performance of biphasic absorption processes by maximizing absorption kinetics and minimizing the concern of high solvent viscosity associated with high CO_2 loading. Totally 34 solvent blends have been formulated and experimentally screened with respect to phase transition behavior, absorption kinetics, and vapor-liquid equilibrium performance. Preliminary results revealed that the phase transition, in terms of its dependence on CO_2 loading and volumetric distribution between two phases, could be tuned with selection of solvent components or composition. The solvents could be formulated to have faster absorption rates than that of the benchmark 5M monoethanolamine (MEA) aqueous solution. The loading capacities of the solvents for CO_2 stripping were 40 to 250% higher than 5M MEA and the stripping pressures generally ranged between 5 and 10 bar under the regenerated lean solution conditions at 120 °C.

In this presentation, an overview of the novel biphasic absorption process will be provided, the preliminary results of the biphasic solvent screening study will be updated, and their indications to the process techno-economic performance will be discussed.

471201 Deterministic and Stochastic Mass Transfer Models for CO_2 Capture Processes

Abstract:

For accelerating deployment of novel CO_2 capture technologies, rigorous process models can be very valuable to reduce the risk and uncertainty in scaling up the process by skipping intermediate scales. For such process models, it is critical that the uncertainties in the key submodels be quantified. For chemical solvent-based post-combustion CO_2 capture technologies, mass transfer models play an important role especially in estimating the performance of the absorber. With this motivation, comprehensive mass transfer models of a chemical solvent based CO_2 capture process with quantified uncertainties is developed in this work.

The typical approach in the literature to development of mass transfer models is sequential. In chemical solvent based CO_2 capture systems such as for the MEA-H_2O-CO_2 system, very fast reaction kinetics make it difficult to obtain measure of the mass transfer flux and identify the associated parameters with the mass transfer models. Assuming the diffusivity model to be satisfactory, mass transfer coefficients are typically calculated from the experimental data from the wetted wall column (WWC) experiments. The experimental data from packed columns are the used to obtain the interfacial area model assuming the mass transfer coefficient models are still valid. As the WWC and packed column operate in different flow and operating regimes, it is not necessary that the mass transfer coefficient models and their parameters are still optimal for the packed column. In addition, assumption of accuracy of the diffusivity model is also questionable. Overall, a suboptimal mass transfer model is obtained. To circumvent these issues, simultaneous regression of parameters for the mass transfer coefficient model, diffusivity model, and interfacial area model is carried out in this work for the MEA-H_2O-CO_2 system by simultaneously using the experimental data from the packed columns and wetted wall columns. Since such simultaneous regression is not feasible in existing state-of-the-art simulation software, a computation tool named Framework for Optimization, Quantification of Uncertainty and Surrogates (FOQUS) developed as part of U.S. DOE's Carbon Capture Simulation Initiative is leveraged to accomplish this objective.

Following the deterministic parameter regression, rigorous uncertainty quantification is performed for the mass transfer models. The best guess for parametric uncertainty (priors) is then cast in the framework of a Bayesian inference methodology to obtain more informed posterior parametric uncertainty. The computational expense for using the Bayesian approach to these models can be prohibitive. Therefore, response surface models are generated as computationally inexpensive surrogate models with sufficient accuracy by using the input-output data. Finally, the posterior distributions in all parameters are propagated through the process model to obtain the uncertainty bounds in key operating variables. Our results show how the uncertainty in process models vary depending on the operating conditions. Our study also indicates possible approaches to reducing uncertainty in mass transfer models.

469614 Adapting Power Generation Lessons Learned to Industrial CO_2 Capture

Abstract:

The Southern States Energy Board (SSEB) is adapting lessons learned in CO_2 capture and storage from power generation to promote commercialization and dissemination of clean energy technologies within the refining and petrochemical industries. The U.S. Department of Energy has partnered with seven Regional Carbon Sequestration Partnerships (RCSP) to advance innovation and research in carbon capture and sequestration. SSEB manages the Southeast Regional Carbon Sequestration Partnership (SECARB), including its Phase III Anthropogenic Test. The Anthropogenic Test is a fully integrated CO_2 capture, transportation, injection, storage and monitoring project that began injecting in August 2012 and concluded injection in September 2014. The project currently is in a 3-year post injection monitoring phase. The post-combustion CO_2 capture facility is a 25 megawatt equivalent advanced amine system, located at Alabama Power Companyâ€™s Plant Barry fossil based power generation facility.

470738 Highly Efficient Warm Gas Carbon Capture System for IGCC Power Plants

Abstract:

TDA is developing a novel sorbent that removes CO_2 via physical adsorption from synthesis gas. The relatively strong affinity of the sorbent to CO_2 enables effective operation at temperatures up to 300°C (well above the dew point of synthesis gas stream generated by commercial gasifiers). However, because the sorbent and the CO_2 do not form a true covalent bond, the energy needed to regenerate TDAâ€™s sorbent (5.4 kcal per mol of CO_2) is much lower than that observed for either chemical absorbents (e.g., 29.9 kcal/mol CO_2 for sodium carbonate) or amine-based solvents (e.g., 14.2 kcal/mol CO_2 for monoethanolamine). TDAâ€™s sorbent can be regenerated isothermally and CO_2 can be recovered at pressure (as high as 150 psia). Thus, the energy needed to regenerate the sorbent and compress the CO_2 for sequestration is significantly lower than that for any other technology reported to date. The high surface area and favorable porosity of the sorbent also provides a unique platform to introduce additional functionality, such as active groups to catalyze the water-gas-shift (WGS) reaction or to remove trace metals (e.g., Hg, As). Hence, TDAâ€™s CO_2 sorbent could be used to do a warm gas capture of CO_2 above the dew point of the synthesis gas in an Integrated Gasification Combined Cycle (IGCC) power plant. TDA's pre-combustion CO_2 capture system achieves a net plant efficiency of 34% on a coal high heating value basis when operated on bituminous coal, which is significantly higher than that can be achieved with the Selexol™ scrubber at 31.6% (state-of-the-art pre-combustion carbon capture technology). The levelized cost of electricity including the transport, storage and monitoring costs for CO_2 is calculated as $92.9/MWh for TDA's system, in comparison to $105.2/MWh for Selexol™. In a current DOE sponsored project, TDA is preparing to carryout field tests with the warm gas CO_2 capture unit at 0.1 MWe scale. In this paper we present the results from the system analysis, bench-scale evaluations, and the fabrication and testing of the 0.1 MWe field test unit.

467400 Development and Optimization of Pressure/Temperature Swing Adsorption for Post-Combustion Carbon Capture

Abstract:

Development and Optimization of Pressure/Temperature Swing Adsorption for Post-Combustion Carbon Capture

K.T. Leperi,[1] F. You,[2] R.Q. Snurr[1]

[1]Northwestern University, Evanston, IL, USA

[2] Cornell University, Ithaca, NY, 14850

With fossil fuels expected to be a significant portion of the world's energy mix for the near future, it is important to minimize CO_2 emissions from current power plants through carbon capture and sequestration (CCS). In post-combustion CCS, the CO_2 is separated from the power plant flue gas emissions, which contain mainly N_2 and CO_2. Among the more promising technologies for accomplishing this separation with minimal energy are adsorption based processes. Only a few studies have focused on temperature swing adsorption (TSA) compared with pressure/vacuum swing adsorption (P/VSA) for separation of CO_2 from N_2. This has been primarily due to long cycle times associated with the heating and cooling of the bed for TSA. However, potential benefits of TSA over PSA are larger adsorbent working capacities for temperature differences compared to pressure differences and the lower cost of using thermal energy to regenerate the bed compared to using electricity for compressors and vacuum pumps. Therefore, it is important to determine not only the optimal regeneration method for CO_2 capture, but also the optimal cycle utilizing these regeneration methods.

In this work, various PSA, TSA and Pressure/Temperature Swing Adsorption (P/TSA) processes are simulated utilizing our previously developed model.[2] This model consists of a system of partial differential algebraic equations incorporating mass and energy balances, pressure drop across the column, competitive multi-site Langmuir isotherms and the linear driving force model. This system of PDAEs is solved by discretizing the spatial dimension utilizing the finite volume method coupled with the weighted essential non-oscillatory (WENO) scheme to generate a system of ODEs, which is solved in MATLAB. The non-dominated sorting genetic algorithm (NSGA-II) is then used to determine the optimal operating parameters for capturing the CO_2 from the flue gas at the lowest cost while maintaining the desired CO_2 purity and recovery. The objective of this work is to compare the different methods for regenerating the adsorbent and identify the ideal cycle for minimizing the cost of capturing CO_2 from flue gas. Each cycle is investigated with zeolites 13X and 5A along with Ni-MOF-74 in order to determine potential influences the adsorbent has on the determination of the ideal cycle and regeneration method.

1. MIT. *The Future of Coal*; 2007. http://web.mit.edu/coal/.

2. Leperi KT, Snurr RQ, You F. Optimization of Two-Stage Pressure/Vacuum Swing Adsorption with Variable Dehydration Level for Postcombustion Carbon Capture. *Ind. Eng. Chem. Res.* 2016;55:3338-3350.

455673 An Investigation of SO_2 Interference on Separation Performance of Amine-Containing Facilitated Transport Membrane for CO_2 Capture from Flue Gas

Abstract:

Post-combustion carbon capture is a key approach to control CO_2 emissions from power plants using fossil fuels. Treating real flue gas may be problematic as it contains not simply just CO_2 and N_2. The minor contaminants, such as SO_2 and O_2, may be the major factors that can limit the long-term stability of a membrane. The typical flue gas after flue gas desulfurization (FGD) treatment is at its discharge temperature of 57°C with 40 – 50 ppm SO_2. The SO_2 level can be further reduced to even 1 ppm by a caustic scrubber, i.e., 20 wt.% NaOH. With the presence of SO_2 in the feed gas, SO_2 could preferably react with amine carriers and hinder the permeation of CO_2 in amine-containing facilitated transport membranes. Possible amine degradation and competitive reactions raise the concerns on the membranes for use in real flue gas. Amine-containing facilitated transport membranes were tested at 102 and 57°C, respectively. SO_2 concentrations of 0.7 – 5 ppm were employed in our transport study. Unstable membrane performance was observed for the membrane tested at 102°C in the presence of SO_2. It was found that amines reacted with SO_2 to form bisulfite and sulfite products irreversibly at 102°C. On the contrary, the membrane showed a stable separation performance in the presence of 1 – 3 ppm SO_2 at 57°C. Moreover, an O_2 concentration of 3% was also added in the feed gas to simulate the real flue gas conditions. The membrane also showed a stable separation performance upon exposure to 3 ppm SO_2 and 3% O_2 for > 100 hours. In addition, infrared spectra were collected with respect to time by using an in-situ FTIR, and the growth of sulfite characteristic peaks was identified. The spectra results were consistent with the observed membrane separation performance, which is an affirmation of the SO_2 interference of the amines in the membrane.

The stable membrane performance at 57°C with SO_2 at ppm levels and O_2 at 3% indicates the applicability of the developed amine-containing membrane for use in CO_2 capture from flue gas.

460857 Engineering Enhanced Permeability and Selectivity in Hybrid CO_2 Capture Membranes

Abstract:

Despite ongoing advances in renewable energy, it remains true that ~90% of global power generation is still derived from carbon-based sources. It is therefore imperative to develop energy-efficient approaches to sequestering CO_2, facilitating decarbonization of energy production. CO_2 capture (pre- or post-combustion) using separations membranes represents an attractive alternative to amine looping, as membranes do not require temperature- or pressure-swings to achieve their performance. Historically, membranes for CO_2 sequestration have been dominated by pure polymers, due to ease of processability and scaling. However, within the solution-diffusion model for pure polymers, it has been challenging to increase CO_2 permeability as molecular diffusion and solubility are typically inversely related. In this talk, I present recent research into developing new molecular transport mechanisms in hybrid organic/inorganic membranes by introduction of chemically modulated inorganic nanocrystals and metal-organic frameworks (MOFs) into classical polymers forming altogether new classes of membranes that offer enhanced solubility, diffusion, and mechanical robustness. I will also briefly discuss strategies for H_2 separations as well.

467335 Two Dimensional Nanosheets-Embedded Thin Film Composite Membrane for CO_2 Cature Applications

Abstract:

The development of cost-effective CO_2 capture technologies for coal power plants will reduce the adverse impact of CO_2 emissions on global climate. Amine absorption technologies have been extensively studied and are regarded viable for post-combustion CO_2 capture. However, the energy comsuption for absorption technologies is still much higher than minimum requirement for large scale post-combustion flue gas. It seems clear that the membrane-based gas separations have presented a suitable alternative to the traditional separation processes for CO_2 capture, particularly for the treatment of flue gas streams including more than 15% CO_2. Membrane systems offer relevant gains in terms of flexibility, reliability, modularity and that they need energy intensity generally lower, or, at least, comparable with that of absorption technology. For competition with existing technologies, most of all, membrane materials with high permeability and high selectivity should be further developed with feasible processibility for real membrane platforms such as hollow fibers or thin-film composite membranes. According to recent report, the membrane systems can compete with existing technologies when membrane shows high CO_2 permeance (~4000 GPU, 1 GPU = 1×10^{-10} $cm^3(STP)/cm^2.sec.cmHg$) and high CO_2/N_2 selectivity (~50). However, to achieve this goal, many factors should be considered: selective layer thickness, gutter layer, and the porosity of support membrane in the form of thin-film composite membrane. In this study, we have developed high permeable, selective thin-film composite membrane for post-combustion CO_2 capture. Selective layers consist of high CO_2 permeable polymer and two-dimensional nanosheets. The nanosheets as dispersed phase in continuous polymer matrix help enhance CO_2/N_2 selectivity (~60) without any loss of CO_2 permeability. Also, such nanosheets at low content help prevent CO_2-induced plasticization, often leading to dramatic decrease in gas mixture. To develop high permeable, selective thin-film composite membranes, we have analyzed the effect of selective layer thickness, types of gutter layer (or gutter layer thickness), and support membrane properties, and these results were used for preparing optimized thin-film composite membranes. These thin-film composite membranes exhibited high CO_2 permeance and high CO_2/N_2 selectivity, even in the presence of water, which is close to the membrane criteria suitable for membrane-based CO_2 capture system. We have also prepared plate-to-frame modules, based on these membranes, showing high CO_2 flux of 5 m_3/h, for pilot plant demonstration.

462467 CO_2 Capture from Power Plant Flue Gas By Polaris^{TM} Membranes: Update on Field Demonstration Tests

Abstract:

To mitigate the harmful effects of global climate change, CO_2 in power plant flue gas must be captured and either sequestered or utilized in processes such as enhanced oil recovery. Membrane technology is an attractive CO_2 capture option because of advantages such as energy-efficient passive operation, tolerance to acid gases and oxygen, no hazardous chemical emissions or handling issues, minimal water requirements, and no steam use requiring modifications to the existing boiler.

Working with the US Department of Energy, MTR has developed new membranes and process designs to recover CO_2 from power plant flue gas. MTR Polaris membranes have CO_2 permeances ten times higher than standard commercial membranes, which greatly reduces the cost of a membrane capture system. These membranes are combined with a novel process design that uses incoming combustion air to sweep membranes and recycle CO_2 to the boiler. Design calculations estimate that this membrane process can capture 90% of the CO_2 in flue gas as a supercritical fluid using approximately 25% of the plant power, at a cost of \$40-\$50/ton of CO_2 captured. This translates to an increase in the levelized cost of electricity of about 50%.

In 2010, MTR completed a three-month field demonstration of the membrane CO_2 capture process with Arizona Public Service (APS) at their Cholla coal-fired power plant. The pilot system used commercial-scale membrane modules, and removed 90% CO_2 from a post-FGD flue gas slipstream containing 1 ton CO_2/day. In late 2011, the system was moved to the National Carbon Capture Center (NCCC) operated by Southern Company in Wilsonville, AL, for further demonstration. Overall, the system has an accumulated run time of over 10,000 hours, and the membrane modules have exhibited stable performance. With experience learned from these tests, MTR has designed a scale-up small pilot unit to capture 20 ton CO_2/day from a power plant slipstream (equivalent to 1 MW_e-scale coal-fired power generation). The system was initially commissioned in August 2014 and testing concluded in June 2015. During testing, advanced modules demonstrating higher packing density, lower pressure drop, and cost savings were validated. In fall 2015, the 20 ton CO_2/day small pilot unit was moved to Babcock & Wilcoxâ€™s (B&W) research facilities in Barberton, OH for a fully integrated recycle test of the small pilot unit with a 1.8 MW_{th} coal-fired research boiler. The field demonstration to be conducted at B&W during the summer of 2016 will be the first time membrane operation with real flue gas and boiler operation with CO_2-laden air have been conducted in a single test. Technical results to date for the field tests, details of an upcoming hybrid membrane-absorption field test at the SRP pilot plant located on the Pickle Research Campus at UT-Austin, and future plans on technical development will be discussed in this presentation.

470281 Multiscale Optimization and Intensification of Natural Gas Separation and Storage

Abstract:

Natural gas is currently the fastest growing energy commodity and the cleanest fossil fuel with abundant reserves. Methane is the predominant component of natural gas. Significant efforts have been undertaken in the past for the separation, storage and transportation of methane. However, the following challenges remain: (i) materials with high selectivity for methane separation from natural gas often do not have high-deliverable capacity for storage, and (ii) materials with high-deliverable capacity for methane storage often do not have high selectivity for separation. In this work, we develop a combined separation and storage (CSS) technology to separate and selectively store methane at the same pressure in the same vessel, thereby essentially eliminating the need for a separate purification process. The proposed method is based on microporous materials (e.g., zeolites). While the state-of-the-art research considers materials for methane separation or storage separately [1], we propose an intensified technology enabled by advanced materials which simultaneously separate and store methane from natural gas.

Using a multi-scale systems engineering approach, we have designed a CSS process that takes a mixture of $CH_4/CO_2/N_2$ as feed, and stores CH_4 inside an adsorbent-packed column while venting N_2 from the system. A non-linear algebraic partial differential equation (NAPDE) model describing the conservation equations and other equations pertaining to adsorption isotherm, mass transfer driving force and pressure drop across the bed is used to study the process. In order to obtain optimal operating conditions, the detailed process model is completely discretized in both space and time into a system of nonlinear equations, whose size increases with the level of discretization. The discrete decisions related to stepwise cyclic operation together with the nonlinear equations result in a large-scale mixed integer nonlinear programming (MINLP) problem. To aid the convergence of the model, simulation-based warm starts are provided to the optimization framework. Optimization is performed to select microporous adsorbents from the database of existing materials [1] and a rank ordered list of candidate materials for different process conditions is generated. Preliminary studies reveal that for achieving simultaneous separation and storage, both purity and storage capacity of CH_4 have to be balanced by identifying the best candidate material and the corresponding operating conditions.

Our multiscale framework has the capability to simultaneously optimize the material and the process. Additionally, we will demonstrate that the framework can be used to obtain optimal properties of hypothetical materials, which can serve as a guide for future experimental efforts to design novel materials for combined gas separation and storage.

References:

[1] First, E. L.; Hasan, M. M. F.; Floudas, C. A. Discovery of Novel Zeolites for Natural Gas Purification through Combined Material Selection and Process Optimization Approach. *AIChE Journal*, 60(5), 1767â€"1785, 2014.

[2] Siderius, D.W., Shen, V.K., Johnson III, R.D. and van Zee, R.D., Eds. â€œNIST/ARPA-E Database of Novel and Emerging Adsorbent Materialsâ€, *NIST Standard Reference Database Number 205*, National Institute of Standards and Technology, Gaithersburg MD, 20899, http://adsorbents.nist.gov.

472821 Ideas Based Synthesis of Fossil Fueled Power Generation/Hydrogen Production Flowsheets with No Carbon Dioxide Emissions

Abstract:

In this work, the Infinite-DimEnsionAl State-space (IDEAS) conceptual framework is employed as a process synthesis activity in the formulation of a flowsheet that utilizes natural gas, carbon, and coal combinations to co-produce hydrogen and formic acid, without any carbon dioxide emissions. IDEAS formulations result in a linear program that allows for a broad search of the state-space in order to identify intensified flowsheets. Employing combustion, steam methane reforming (SMR), pressure swing adsorption, and formic acid production from CO2 and H2, a carbon dioxide emission free flowsheet can be synthesized. Thermodynamic analysis, heat and power integration studies, and operating costs/revenue studies are all presented to determine the designâ€™s operating revenue to cost ratio.

467387 A Heuristic-Based Model for Gasification-Based Plants with Flexible Feedstock/Product in Carbon Constrained Scenarios

Abstract:

Gasification is one of the most versatile thermochemical conversion technologies for carbonaceous feedstocks. The inherent feedstock flexibility of gasification technology as well as the broad range of end use products makes it an attractive option in coal and biomass conversion. The commercially available technologies for pre-combustion CO_2 capture from syngas are extremely capital intensive. The economic impact of a carbon tax on a gasification based plant significantly depends on the quality of feed, technology, and application of interest.

In this study, the impact of the sulfur, ash, moisture content, and heating value are considered for a wide range of coals and biomass. Heuristics, continuous and discrete design variables and cost functions are used to build various modules for major process units including gasification island, acid gas removal, water gas shift reactor, and application specific processes. A heuristic-based algorithm is used to identify the near-optimum modules (nodes) of gasifier and downstream subsystems to determine the process route with the maximum net present value under various carbon tax schemes for a given set of inputs and constraints.

This model examines a myriad of alternative pathways to determine the most economical match of carbonaceous feedstock composition, gasifier choice, and syngas processing subsystems to a process application of interest including electricity, hydrogen, methanol and Fischer-Tropsch hydrocarbon production.

470303 Thermodynamic Modeling and Uncertainty Quantification of CO_2-Loaded Aqueous MEA Solutions

Abstract:

The US DOE's Carbon Capture Simulation Initiative (CCSI) has a strong focus on the development of state of the art process models to accelerate the development and commercialization of post-combustion carbon capture system technologies. One of CCSI's goals is the development of a "gold standard" process model that will serve as a definitive reference for benchmarking the performance of MEA-based CO_2 capture systems. The focus of this specific work is the development of the thermodynamic framework, an essential element of the physical properties package for a reactive separation system, for the gold standard model. A stochastic model of the thermodynamic framework is also included in this work, in which a Bayesian inference methodology is used to quantify the parametric uncertainty.

The thermodynamic framework is developed in Aspen Plus®, and the complete e-NRTL activity coefficient model is used as a starting point. The Aspen Plus® physical properties regression system is used to calibrate the parameter values, including data for VLE of binary MEA-H_2O and ternary MEA-H_2O-CO_2 systems as well as heat of absorption and heat capacity of the ternary system. Due to the large number of parameters in the full e-NRTL model, a parameter screening methodology based on the Akaike Information Criterion (AIC) is developed. The use of a reduced model is not only useful in avoiding over parameterization in the deterministic model, but also improves the feasibility of the uncertainty quantification (UQ) procedure. A Bayesian inference methodology is used to develop the stochastic model, in which prior distributions estimated from the deterministic regression output and the experimental data are used to derive informative posterior distributions. To reduce the computational expense of the Bayesian approach, response surface models are generated as computationally inexpensive surrogate models with sufficient accuracy by using the input-output data. The Markov Chain Monte Carlo (MCMC) algorithm with Gibbs Sampling is used in the Bayesian inference procedure.

The thermodynamic framework is implemented into an Aspen Plus® model of the CO_2 absorption process, consisting of both absorber and regenerator columns, which are modeled as rate-based columns with kinetic reactions. The reaction kinetics used in this work are written in terms of the activity-based reaction equilibrium constants in order to ensure consistency with the thermodynamic model. Case studies are performed, separately for the absorber and regenerator columns, in which prior and posterior parameter distributions are propagated through the process models. This results in estimates of the uncertainty in output variables, specifically CO_2 capture efficiency in the absorber and outlet solvent CO_2 loading from the regenerator, due to the thermodynamic model. The study shows that the uncertainty due to the VLE model can vary widely depending on the operating conditions, but does reduce considerably for the posterior distributions in comparison to the priors for all the cases studied. The study also helps to identify how the uncertainty due to the thermodynamic model can be reduced further.

467832 Design and Advanced Optimization of a Natural Gas Combined Cycle Power Plant with CO_2 Capture

Abstract:

As the penetration of intermittent renewable energy sources into the electrical grid keeps increasing, traditional fossil-fired power plants would be expected to cycle their load much more frequently than in the past. However, current fossil-fueled generation plants are not optimally designed for cycling. With this motivation, a rigorous model of a natural gas combined cycle (NGCC) plant with CO_2 capture is developed. The plant is then designed for optimal load-following characteristics.

First, a steady-state model of the NGCC plant with a triple-pressure reheat cycle steam generator is developed in Aspen Plus. A stage-by-stage model of the steam turbine is developed in Fortran and implemented as a user model in Aspen Plus for estimating the performance of the triple-pressure steam turbine with multiple steam addition and extraction points including the large LP steam extraction for the CO_2 capture process. A mixture of piperazine-promoted methyldiethanolamine is used as the solvent for CO_2 capture. A rate-based model for this solvent system is developed and validated using available experimental data. While traditional optimization approaches in this area have looked mainly into maximizing efficiency, this work is focused on optimizing a cost function that considers both the operating costs as well as the capital costs required for improved load-following characteristics. As an example, even though increased steam temperature and pressure increases the overall efficiency, it increases steam-side oxidation of the boiler tubes especially the high pressure superheater tubes resulting in lower thermal conductivity, which leads to higher skin temperatures. This effect not only leads to higher corrosion for the tubes but also results in lower strength of the tube materials thereby necessitating a design with greater tube thickness. The oxide scale also has a different thermal expansion coefficient compared to the base material leading to stress buildup that can result in spalling especially during startup and shutdown of the plant. Furthermore, creep and fatigue characteristics of the plant under design conditions, during operations under elevated conditions, and under thermal cycling should be optimally designed for load-following. An evolutionary optimization technique is setup by bridging optimization software with Aspen Plus to maximize the net present value. Pareto curves in the presence and absence of the CO_2 capture process are generated showing the trade-offs between the operating and capital costs.

462335 Natural Gas to Liquid Transportation Fuels Utilizing Chemical Looping Technologies for Syngas Generation: Process Synthesis and Global Optimization

Abstract:

Even with advances in electric vehicles and improved vehicle fuel economy, the demand for liquid fuels in the transportation sector is projected to remain steady through 2040 [1]. Liquid fuels produced from domestic feedstocks would partially offset this demand and could ultimately reduce dependence on foreign crude imports, enhancing national energy independence. A potential feedstock is natural gas due to its high hydrogen to carbon ratio and methane-rich composition, increasing the overall carbon conversion to liquid fuels instead of CO_2. Natural gas to liquid fuels (GTL) processes may also be economically promising due to an abundance of inexpensive natural gas in the United States (2015 spot price: $2.62/MMBtu [2]).

Previous multi-scale engineering [3] work by Baliban et al. [4] demonstrated that optimal GTL processes can be economically competitive with petroleum refineries. The primary natural gas conversion technologies investigated were syngas generation via autothermal reforming or stream reforming, partial oxidation to methanol, and direct oxidative coupling to olefins, before further upgrading to liquid fuels via methanol synthesis or Fischer-Tropsch processes. Natural gas conversion is low through partial oxidation and oxidative coupling routes, and while it is greater through traditional reforming, high pressure operation limits the overall conversion.

However, recent developments in chemical looping [5, 6] as an alternative for syngas generation from natural gas have significantly expanded the potential GTL capabilities. Chemical looping can be operated at low pressure and offers close to complete conversion of natural gas in a single pass. A high concentration of syngas can be produced without using pure oxygen, eliminating the need for additional air separation and syngas conditioning units. This could greatly reduce the capital and operational costs associated with generating syngas in GTL processes and improve the overall process efficiency.

In this work, two chemical looping technologies are incorporated into a GTL process superstructure [4, 7, 8, 9] as alternatives to reforming for syngas generation. All process technologies are rigorously modeled and together with other important process components, such as hydrogen and oxygen production, wastewater treatment, and light gas handling, form a large-scale nonconvex mixed-integer nonlinear model (MINLP). A deterministic global optimization branch-and-bound algorithm is used to solve the MINLP [10]. Simultaneous heat and power integration is also performed to minimize utility cost [11]. This process synthesis framework provides an illuminating means to systematically compare chemical looping technologies against other competing technologies in a GTL process.

Several case studies are examined to highlight the prospective benefits of chemical looping technologies over other conversion routes. The effect of plant capacity and production ratios on the overall profit of the GTL process is analyzed. Major topological decisions on process technologies will be discussed. Economic and environmental trade-offs will also be presented.

[1] U.S. Energy Information Administration. Annual Energy Outlook 2015 with projections to 2040. Available at: https://www.eia.gov/forecasts/aeo/pdf/0383%282015%29.pdf. Accessed May 2016.

[2] U.S. Energy Information Administration. "Henry Hub Natural Gas Spot Price". Available at: http://www.eia.gov/dnav/ng/hist/rngwhhdm.htm. Accessed May 2016.

[3] Floudas, C. A.; Niziolek, A. M.; Onel, O.; Matthews, L. R. Multi-scale systems engineering for energy and the environment: Challenges and opportunities. AIChE Journal 2016, 62 (3), 602-623.

[4] Baliban, R. C.; Elia, J. A.; Floudas, C. A. Novel Natural Gas to Liquids Processes: Process Synthesis and Global Optimization Strategies. AIChE Journal 2013, 59 (2), 505-531.

[5] Luo, S.; Zeng, L.; Xu, D.; Kathe, M.; Chung, E.; Deshpande, N.; Qin, L.; Majumder, A.; Hsieh, T.-L.; Tong, A.; Sun, Z.; Fan, L.-S. Shale gas-to-syngas chemical looping process for stable shale gas conversion to high purity syngas with a H2:CO ratio of 2:1. Energy & Environmental Science 2014, 7 (12), 4104–4117.

[6] de Diego, L. F.; Ortiz, M.; García-Labiano, F.; Adánez, J.; Abad, A.; Gayán, P. Hydrogen production by chemical-looping reforming in a circulating fluidized bed reactor using Ni-based oxygen carriers. Journal of Power Sources 2009, 192 (1), 27–34.

[7] Baliban, R. C.; Elia, J. A.; Floudas, C. A. Biomass and Natural Gas to Liquid Transportation Fuels: Process Synthesis, Global Optimization, and Topology Analysis. Industrial & Engineering Chemistry Research 2013, 52 (9), 3381-3406.

[8] Onel, O.; Niziolek, A. M.; Elia, J. A.; Baliban, R. C.; Floudas, C. A. Biomass and Natural Gas to Liquid Transportation Fuels and Olefins (BGTL+C2_C4): Process Synthesis and Global Optimization. Industrial & Engineering Chemistry Research 2015, 54 (1), 359-385.

[9] Niziolek, A. M.; Onel, O.; Elia, J. A.; Baliban, R. C.; Floudas, C. A. Coproduction of liquid transportation fuels and C6_C8 aromatics from biomass and natural gas. AIChE Journal 2015, 61 (3), 831-856.

[10] Baliban, R. C; Elia, J. A; Misener, R.; Floudas, C. A Global optimization of a MINLP process synthesis model for thermochemical based conversion of hybrid coal, biomass, and natural gas to liquid fuels. Computers & Chemical Engineering 2012, 42, 64-86.

[11] Baliban, R. C.; Elia, J. A.; Floudas, C. A. Simultaneous process synthesis, heat, power, and water integration of thermochemical hybrid biomass, coal, and natural gas facilities. Computers & Chemical Engineering 2012, 37, 297-327.

470988 Optimal Design and Operation of Hybrid CO2 Capture Systems

Abstract:

This work presents a mixed integer nonlinear programming model (MINLP) for the optimal design and operation of carbon capture plants. A mix of first principle and surrogate-based models have been developed to characterize the process design and operation.

Coal-based power generation is one of the most important sources of energy in the US. As such, the importance of cost effective carbon capture and storage (CCS) technologies to mitigate the environmental issues of power generation plants is obvious. The most significant technologies for carbon capture include solvents, membrane and solid sorbent systems. This work analyzes and compares the solid adsorption and membrane systems on the basis of minimizing the cost of electricity.

In solid sorbent capture systems the process is mainly driven by the interaction of the solid sorbent particles with the flue gas (mainly consisting of N_2, CO_2, and H_2O). The main process units of the solid sorbent systems are multiphase reactors, such as moving bed and fluidized beds.

In gas permeation separation systems, the CO_2 is separated by high CO_2 selective membranes. The main driving force of the process is the partial pressure between the feed and the permeate side. A typical membrane separation plant consists of multi-stage membrane configurations and compressors.

Given the modeling complexity and the difficulty of formulating algebraic models for such processes, previous studies are generally limited to either process-simulation[1] based frameworks or simpler models for mathematical optimization. This work proposes a mathematical optimization framework in which both first principles-based models and surrogate models[2] have been developed to obtain the optimal plant design and operation to capture at least 90% of the CO_2. The use of commercial process simulators to develop detailed, carefully tuned models have been exploited in the preparation and characterization of surrogate models (ALAMO[2]). Model validation and sample generation play a very important role in this mathematical framework.

The modeling framework considers a set of adsorbers (with two possible operating technologies), regenerators (with two possible operating technologies), membrane stages, mixers, splitters, compressors, pumps and heat exchanger units which can be installed or not. Design decisions determine the plant layout, by selecting the optimal number of parallel trains, number of adsorbers and regenerators, number of membranes, equipment design (diameter, length, etc.), heat exchangers, mixers, splitters and configuration of the system. At the operating level, the decisions involve the optimal flows, temperatures, concentrations of gas and solids. For the adsorbers and regenerators, the solids bed length, flux velocity and other parameters are also considered. While, the membrane characteristics are to be determined by the permeability, permeate fraction, pressure ratio, among others.

[1] Eslick, J. C., Tong C., Lee A., Dowling A. W., Mebane D. S. (2015). Optimization under uncertainty with rigorous process models. Proceedings AIChE Annual Meeting 2015.

[2] Cozad, A., N. V. Sahinidis and D. C. Miller, Learning surrogate models for simulation-based optimization, AIChE Journal, 2211-2227, 2014.

464491 An Investigation into the Effects of Impurities in Captured CO_2 stream on Safety and Environmental Indicators within Transportation Pipeline and Downstream Storage

Abstract:

Abstract

The realms of carbon dioxide pipeline transportation and its subsequent storage have long been stirring many incentives for multidisciplinary research across the world. However, there are several aspects of these processes yet to be explored in detail. Amongst these, the coexistence of certain impurities in transportation pipeline and/or storage facilities along with carbon dioxide is of utmost significance.

In our previous works we defined the indices corresponding to the safe operation of transport and storage processes with regard to the concentration of specific impurities (factors). Along with these impurities, three safety indices (responses) were considered i.e. pipeline corrosion, acute and chronic toxicities. However, these studies were focused on the effects of individual impurities alone and lacked any actual combined effects beyond additive ones. Therefore, little has been established on how these so-called impurities would affect these indices when present simultaneously within such processes.

In this study, the interaction effects of several impurities including water, oxygen, NO_2, SO_2, Cl and several heavy metals (i.e. Pb, Hg, Cd and As) on the above safety indices have been investigated in order to define the resulting safe operational envelopes. The interactions were revealed by initially developing a full factorial DoE framework and then implementing an additive approach to represent the combined indices in each run. These interaction were next amalgamated in order to produce the corresponding 3D and contour plots which highlight the boundaries between the safe and unsafe operating conditions.

This study sheds further light on how the combined effects of various impurities present in the captured CO_2 stream would affect the transportation as well as the downstream infrastructures. This is expected to potentially bring about complementary considerations. A further study to bridge between the identified safe operating conditions and the capture cost is now under investigation. This would aim to successfully provide the decision makers and engineers with a better understanding of the associated costs linked to the required safety considerations as well as the underlying trade-offs.

474284 Natural Gas Upgrading Towards Value-Added Chemicals (keynote)

Abstract:

In the United States and around the world, there are abundant gas resources which are either physically or economically stranded. These gases include flaring gas in shale oil field, refinery off-gas, coal-bed methane, shale gas in disadvantaged locations, etc. Upgrading of these gases to value-added liquid products could reduce the demand on crude oil by about 20% in the United States. Indirect conversion of natural gas to liquid products (GTL) via syngas has been commercialized but it requires huge capital investment, varying from $15 to 20 billion for a 100-140 bbl/d plant. Direct conversion of natural gas without going through syngas route has been under investigation in the past 5 decades, but no commercial processes are practiced to date. A large number of studies have been published on the subject over the past 50 years. This presentation illustrates experimental research on direct non-oxidative conversion of natural gas. The challenge in advance the fundamental science aspects presented in direct methane conversion is discussed. By comparing with propane dehydrogenation technology (Catofin), the industrial perspective on direct methane conversion is illustrated. Specifically, the strategies on the selection of reactor configuration, heat management and catalyst regeneration are highlighted. These commercialization strategies along with technoeconomic analysis are critical for direct natural gas conversion to value-added chemicals.

471978 Methane Dry Reforming over Rh and Ru Based Pyrochlore

Abstract:

Dry reforming of methane was studied over LSZ pyrochlore catalysts containing Rh and Ru. The 1%LSRZ and 1%LSRuZ catalysts were synthesized using a variation of Pechini method. Both the catalysts have a Lanthanum Strontium Zirconate support with Rhodium or Ruthenium substitution. The catalysts were characterized by TPR, XRD, and ICP. Five cycles of TPR's were carried out for each catalyst with intermediate TPOâ€™s. The results showed an increase in reducibility for both the catalysts from one cycle to another. The catalysts were tested for their dry-reforming activities at different temperatures. Long time (90 hrs) testing of the catalysts were carried out at a reference temperature of 700 Â°C. The results showed that the Ru-substituted catalyst deactivated quickly, whereas the Rh-substituted catalyst showed practically no deactivation with time-on-stream. Post-reaction TPO for the catalysts showed carbon deposition in various forms â€“ amorphous, polymeric, and graphitic. Even though 1%LSRuZ shows deactivation with time-on-stream, the catalyst exhibited lesser carbon deposition on the surface. This work clearly shows the effect of metal on the progress of the reaction and Rh metal was thus identified to be a better active metal for the dry reforming reaction.

470681 Upgrading Methane to Olefins Using Oxidative Coupling of Methane in a Chemical Looping System

Abstract:

Methane is viewed as a fuel or an energy source, with limited applications in its direct use as a feedstock for chemicals manufacturing. The high molecular stability of methane requires an indirect approach where it is first converted to an intermediate chemical like syngas, which can then be used for producing commodity chemicals. This method of methane conversion requires multiple processing steps and are highly energy intensive. The direct approach for methane conversion to value-added products, on the other hand, allows for process intensification, which makes it competitive both in terms of cost and energy. Oxidative Coupling of Methane (OCM) has the highest per pass yields towards products among all existing direct methane conversion technologies. It also provides a one-step process towards manufacturing olefins (C_{2+} hydrocarbons) from methane. Traditionally, methane oxidation in OCM has been carried out by co-feeding molecular oxygen with methane over an activated catalyst. The chemical looping method, however, utilizes the lattice oxygen in a metal oxide catalyst for methane oxidation, which results in a higher selectivity towards C_{2+}as compared to the results from the traditional co-feed process. The intended use of the olefinic product stream could be to separate it into different components and integrate it with existing refinery operations. In another approach the product stream can be sent to an oligomerization reactor to convert the unsaturated hydrocarbons to gasoline over a ZSM-5 catalyst. Such a process shows flexibility in its utilization as a commercial technology.

A manganese-based oxide was synthesized and used as the metal oxide catalyst for this study. Gases other than pure methane like shale gas and biogas were also used to demonstrate the versatility of the catalyst. Parameters such as reaction pressure, steam flow rate and gas-hourly space velocities were varied to observe their effect on the amount of C_{2+} products produced. These experiments were performed in a fixed bed condition, cycling between oxidation and reduction steps. The results obtained were coupled with Thermogravimetric Analysis (TGA) of the metal oxide catalyst

to investigate recyclability of the catalyst when subjected to multiple cycles. The effects of these parameters was also analyzed by nitrogen physisorption (BET) for pore volume and surface area changes from the original metal oxide catalyst. The studies performed were directed towards the application of the manganese based catalyst in a co-current moving bed reactor.

461994 Chemical Looping - Oxidative Dehydrogenation for the Green Production of Ethylene

Abstract:

Ethylene is an important precursor for a variety of materials used today. The traditional method for ethylene production, i.e. steam cracking of hydrocarbons, is highly energy intensive due to the endothermic nature of the reaction, the high process temperatures, and heavy separation loads for downstream purification. The generation of the energy necessary for this process leads to an annual emission of over 150 million tonnes of CO_2 worldwide. Oxidative dehydrogenation of ethane (ODH) has been investigated as a potentially more efficient approach. In ODH, ethane is partially oxidized into ethylene and water. The oxidation of hydrogen makes the ODH reaction exothermic, thereby reducing fuel needs while overcoming the equilibrium limitations for cracking reactions. However, to allow economical product separation, conventional ODH schemes require the use of pure oxygen co-fed with ethane. The oxygen generation, which uses cryogenic air separation systems, are capital and energy intensive.

We propose a chemical-looping ODH (CL-ODH) approach that can address these issues. In this scheme, a metal oxide based redox catalyst provides oxygen for the ODH reaction from its lattice. The oxygen depleted redox catalyst is subsequently oxidized in air, regenerating the catalyst and releasing the heat needed for the process. In this work we show proof of concept and characterization data for a model CL-ODH redox catalyst. Additionally, we report the results of a process modeling study based upon preliminary reaction data. It is shown that an active CL-ODH catalyst can be made based upon a doped Mg_6MnO_8 system. Simulations indicate that CL-ODH has the potential to reduce CO_2 and NO_x emissions by over 80%, and that system is projected to be more energy efficient than steam cracking even when the fuel value of byproducts is considered.

452400 Design and Properties of Supported Tungstate Catalysts for Oxidative Coupling of Methane (OCM)

Abstract:

Methane is the largest component of natural gas and possesses numerous thermodynamic and kinetic challenges in its selective one-step conversion to value-added products. At the same time, creating one-step catalytic processes is of critical importance in monetizing natural gas due to the remote locations of extraction sites and high transportation costs, but no economically viable process exists. Current *state of the art* involves multi-step capital intensive syngas production, water-gas shift to control the CO/H_2 followed by Fischer Tropsch or methanol synthesis. We present on the structure and property of optimized WO_x catalytic sites supported on SiO_2 promoted with Na and Mn to improve current yield of C_2 products during OCM. We performed synthesis, molecular and electronic structural characterization under reaction conditions (*operando* Raman and Density Functional Theory) on the current most selective catalyst, supported $Mn/Na_2WO_4/SiO_2$. These studies isolated the catalytic active site(s), determined their molecular/electronic structures and establish their fundamental structure-activity relationships. These combined experimental and theoretical findings can be used to guide the rational design of advanced OCM catalysts by tuning of the catalytic active site(s) to molecular engineer improved activity and selectivity performance. The new insights that will be presented will lead to the development of conceptually new catalyst models that will have the potential to enable development of CH_4 single-step OCM catalytic processes using a bottoms up approach via catalytic active site(s), rather than top down approach of catalytic screening experiments.

459773 Water Splitting and Partial Oxidation of Methane Using an Oxygen Permeable Membrane with Ni Catalyst

Abstract:

In this presentation, we will discuss hydrogen production from water thermolysis and partial oxidation of methane (POM) in an oxygen permeable MIEC (mixed ionic-electronic conducting) membrane reactor. Water thermolysis on the feed side is the oxygen source, and oxygen diffuses through the membrane to react with methane on the sweep side. Conventional nickel catalysts for POM suffer from carbon deposition due to the insufficient supplies of oxygen. Besides, due to the exothermic reaction of POM, the catalysts experience high temperature spots, which lead to melting or agglomeration. These two problems can be solved in an oxygen permeable membrane reactor with water thermolysis.

Oxygen permeable $La_{0.9}Ca_{0.1}FeO_{3-\delta}$ (LCF-91) membrane is used [1] [2]. The overall reaction is endothermic, so hot spots on nickel catalysts can be avoided. Moreover, carbon deposition from methane decomposition is suppressed by increasing the oxygen flux through the application of nickel catalysts. Experimental results show that optimum syngas composition for gas-to-liquid pathway i.e., $H_2/CO = 2$ is achieved on the sweep side. Local measurements on the membrane surface reveal that the methane oxidation mechanism depends on the availability of surface oxygen. The catalytic porous layer on the sweep side also enhances the water thermolysis by two orders of magnitude. Long term study shows that the catalytic layer maintains high performances for 54 hours.

Additionally, a novel approach is adapted to synthesis nickel nanoparticles on perovskite supports on the sweep side. Because of thermodynamic equilibrium, nickel atoms exsolve from perovskite lattices and form nanoparticles on the perovskite grains with good distribution under reducing conditions. These nickel nanoparticles are expected to have high catalytic performances and good stability; and hydrogen production from water and methane in the oxygen permeable membrane can be enhanced.

Reference:

[1] Wu, X. Y., Chang, L., Uddi, M., Kirchen, P., and Ghoniem, A. F., 2015, "Toward enhanced hydrogen generation from water using oxygen permeating LCF membranes," PCCP, 17(15), pp. 10093-10107.

[2] Wu, X. Y., Uddi, M., and Ghoniem, A. F., 2016, "Enhancing co-production of H2 and syngas via water splitting and POM on surface-modified oxygen permeable membranes," invited to submit to AIChE Journal, under review.

474898 Research & Development Needed to Enable Continued Use of Coal in a Carbon-Constrained Future

Abstract:

Earlier this year more than 100 nations signed the COP-21 agreement which calls for CO_2 levels in the atmosphere to be limited to no greater than what would lead to a 2Â°C increase in the global average temperatures when compared to the pre-industrial era. This would require a significant reduction in CO2 emissions. A reduction which many researchers have said will eventually require the use of CO_2 capture and storage (CCS) or utilization on almost all fossil fuel power plants. However, todayâ€™s CCS technology is expensive and is unlikely to be applied unless the price for emitting CO2 exceeds $60 to $100/ton. Moreover, in many locations neither geologic storage nor enhanced oil recovery via CO_2 injection is available, and the applicability of utilization is also limited.

Nevertheless, there are options for decreasing emissions from coal power plants which do not involve using CCS. These options are aimed at increasing the thermal efficiency of coal power technology, so that less coal is used to generate a given amount of power.

The Electric Power Research Institute (EPRI) has conducted an analysis which shows that some advanced coal-based technologies could achieve a CO2 emissions standard of 500 to 636 kg/MWh. Several of these technologies could be applicable for retrofitting to todayâ€™s fleet of coal power plants. This keynote will highlight some of the technologies which could decrease CO2 emissions from coal power plants. It will also identify the most important research and development (R&D) which must be undertaken to bring those technologies to commercial reality, and will argue that R&D on CCS must continue as CCS is the only technology which provides the level of CO2 reductions required to meet the COP-21 target.

474720 Efficiency and Reliability of a Coal Fired Power Plant in the Age of Increased Renewable Power Supply

Abstract:

Renewable energy contribution to the generation of electric power continues to increase while the overall contribution of coal fired generation is steadily decreasing. The daily and hourly fluctuations of renewable energy adds increased pressure on the coal fired units to produce electricity reliably and efficiently. It has been shown that during load transient conditions both plant efficiency and equipment material life decrease. Coal fired units operating in a stand-by condition can see up to a 10% reduction in their efficiency (i.e., an equivalent increase in CO_2 emissions). Equipment reliability could be impacted by increased cycling due to fatigue failures and other phenomena such as exfoliation in certain metals. The discussion will review areas where additional Research & Development may be needed to improve the reliability and efficiency of these plants during transient and partial load conditions.

458377 Advances in Autothermal Reformer Development

Abstract:

Advances in autothermal reformer development

At the Forschungszentrum Jülich, at the Institute of Energy and Climate Research (IEK-3), intensive research and development in the field of reactors for autothermal reforming (ATR) of diesel fuel and kerosene was performed within the last 15 years. It is described in a number of scientific papers. Most reactors have in common that they are equipped with a pressure swirl nozzle for cold fuel injection. They all have a fuel evaporation chamber, in which fuel is vaporized and mixed with steam, as well as an air mixing area, in which air is injected and blended with steam and evaporated fuel. The core component of any of Jülich's autothermal reforming reactors is the catalyst. In most cases, it is a bimetallic RhPt species, which is coated on an Al_2O_3-CeO_2 washcoat, which in turn is deposited on a monolithic cordierite substrate. Steam is generated in an internal device for heat exchange using the waste heat from the product gas flow of the autothermal reforming reaction. Figure 1 illustrates these common characteristics of Jülich's ATRs.

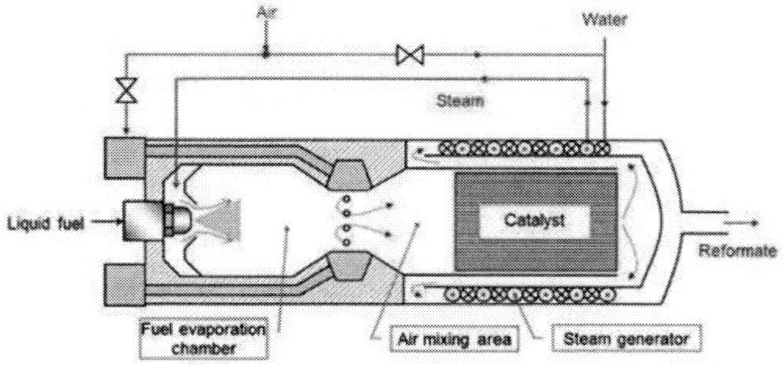

Figure 1 Basic layout of Jülich's reactors for autothermal reforming of diesel fuel and kerosene

This contribution deals with Jülich's recent technical and scientific advance in the field of reactors for autothermal reforming. This reactor generation is denoted as ATR 12.

ATR 12 is a consistent progression of ATR AH2, which was recently published in [1]. It is reported in this paper that the improvements of ATR AH2 in comparison to former ATR reactor generations from Jülich consisted of an additional pressure swirl nozzle for the injection of cold water and a steam generation chamber. As a result, no external process configuration for steam supply was necessary any more. Additionally, the cross-sectional area of the steam generator in ATR AH2, through which also a significant air flow is fed to the reformer, was increased, while in parallel its length was reduced. Thereby, the pressure drop of the steam generator drastically decreased. These modifications in the autothermal reformer design are not decisive for the stand-alone operation of the reformer, but very beneficial from the fuel cell system perspective since they lower the parasitic losses caused by the compressor for air supply and make the system layout simpler. For ATR 12, this way to consider important technical requirements for the design and construction being imposed by the fuel cell system was continued. For ATR 12, the technical execution of the internal steam generator was substantially changed and further improved. The coiled tubing for transferring the waste heat from the product gas flow of autothermal reforming to the stream of saturated steam coming from the catalytic burner of the fuel cell system was substituted by concentric shells, in which turbulence inserts were placed to improve the heat exchange properties of the new construction. This modification aims at an additional decrease in the pressure drop in this part of the reactor and a homogenization of the media streams, when they enter the fuel evaporation chamber. Apart from the above mentioned turbulence inserts, the concentric shells also offer enough space for the incorporation of an electric heating wire. This modification provides the option of fast and autonomous start-up of the autothermal reformer. In this respect, Figure 2 shows the specific design of Jülich's recent reactor type for autothermal reforming ATR 12.

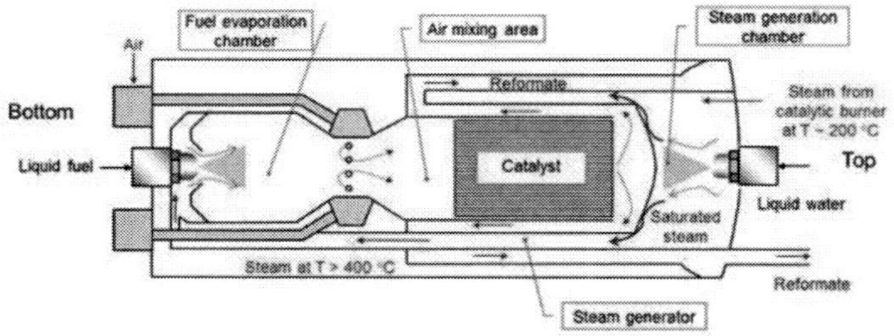

Figure 2 Specific design of Jülich's recent reactor type for autothermal reforming ATR 12

Three different experimental procedures for fast heat up of ATR 12 and additional steady-state experiments are presented in this contribution. Steady-state experiments aimed at validating the concept for heat management of ATR 12. Critical temperatures (internal steam, fuel evaporation chamber, air mixing area, to water-gas shift reactor) and product gas concentrations (H_2, CO, CO_2, CH_4, ethane, ethene, propene and benzene) of ATR 12 are presented as a function of the reformer load and the mass fraction of cold water to the nozzle on the top side of the reformer.

[1] J. Pasel, R.C. Samsun, A. Tschauder, R. Peters, D. Stolten, A novel reactor type for autothermal reforming of diesel fuel and kerosene, Applied Energy 150 (2015) 176-184.

453117 Stable Start-up Process for 1 kW_e Diesel Autothermal Reformer for Auxiliary Power Unit Applications

Abstract:

Nowadays, Auxiliary Power Units (APUs) have a great potential to provide distributed power for remote regions. Also, APU can be used for mobile applications such as heavy duty trucks. Energy density is an important factor for fuels for the APU applications due to its limited space; volume and weight are restricted to ensure mobility. In that manner, diesel can be one of the good candidate fuels for the APU applications due to its high energy density.

However, typical diesel engine generator shows low electric conversion efficiency and makes high pollution. Therefore, a Solid Oxide Fuel Cell (SOFC) based APU is suggested to generate electricity, providing higher efficiency and lower pollutant emission. To utilize SOFC for the APU applications with diesel fuel, high quality hydrogen feed should be provided. With autothermal reforming (ATR) reaction of diesel fuel, hydrogen-rich gas can be produced with catalytic reactors.

During the operation of the diesel ATR, stable start-up process is a critical factor to determine its long-term stability. Initial degradation can severely occur if improper start-up process is applied, and the degradation affects to the durability of the diesel reformer. Also, start-up time must be reduced to minimum to provide fast electricity generation. Therefore, it is needed to develop stable and fast start-up process for the diesel autothermal reformer.

In this study, start-up strategies for diesel autothermal reformer are investigated. With initial heat-up process with electrical heater, partial oxidation mode boosts the start-up process. Temperature distributions within the reformer parts are monitored to manipulate inlet flow conditions. As the result, start-up strategy which enables less than 30 minutes of start-up time is obtained for the 1 kW_e diesel autothermal reformer.

467034 Production of Hydrogen Via Partial Dehydrogenation of Fuels

Abstract:

Production of Hydrogen Via Partial Dehydrogenation of Fuels, to Feed a Fuel Cell

Mélanie Taillades, Julien Belloc, Gilles Taillades, Deborah Jones, Jacques Rozière

Institut Charles Gerhardt UMR 5253, Agrégats, Interfaces et Matériaux pour l'Energie,

Université de Montpellier, Place Eugène Bataillon, 34095 Montpellier Cedex 5, France

One of the objectives of the industries is to provide more efficient and greener transport environment [1]. In this context a new generation of more electrified vehicles must be developed. One of the technologies on this roadmap is a fuel cell secondary power generation unit replacing the current Auxiliary Power Units (APU) [2-4] and proton exchange membrane fuel cells (PEMFC) are the best candidates for the role as an APU replacement. PEMFC operate on pure hydrogen and air between 50 to 100 °C (120 to 200 °C for high temperature PEMFC), and produce electricity and water, with heat and oxygen-depleted air as by-products

The two main issues for the actuation of such an energy system based on hydrogen are its production and storage. Hydrogen production mainly relies on fossil fuels and application to the transport sector would mean a big increase of the demand and hydrogen storage has the drawback of weight, overall storage size and security. This means a big challenge to find a safe and economic way to store it and make it available for applications of interest, in particular for transportation on-board application.

Among the different processes for the hydrogen delivery to fuel cells, a promising one is the generation of hydrogen by catalytic dehydrogenation of liquid hydrocarbons (HCs). This seems to be an appropriate solution as it can produce the hydrogen directly from vehicle fuel, but is only feasible if the process is able to produce sufficient pure hydrogen.

Ours works are based on the concept for hydrogen generation by partial dehydrogenation (PDh) of fuels, allowing the production of dehydrogenated hydrocarbons in liquid phase and hydrogen in gaz phase [5, 6, 7]. Hydrogen has high purity and is CO_2 free, allowing the direct alimentation of fuel cells for supply of electrical energy to auxiliary systems, without a purification step. The liquid phase, composed of partially dehydrogenated hydrocarbons, maintains its original fuel properties with the requisite specifications to be reused as fuel. This process allows the production of hydrogen either on board with the PDh of the vehicle fuel or in a generator connected to the existing fuel storage facilities. The modified fuel can be re-injected into the pool of the vehicle or in fuel storage existing facilities.

This work describes research works on the partial dehydrogenation of gasoline for the production of enough pure hydrogen to feed a fuel cell on board a car. The choice of the catalyst is crucial for the partial dehydrogenation process; it must produce H_2 without compromising the original fuel properties. An ideal catalyst must generate sufficient hydrogen of high purity, be selective to dehydrogenation and avoid cracking reactions responsible for coke deposition and catalyst deactivation. Advanced material, composed by an active phase of 1% Pt-1 % Sn impregnated on a new γ-alumina type support has been developed, characterized and evaluated as a catalyst in the reaction of partial dehydrogenation. The influence of catalyst composition on the activity, selectivity and stability as well as a regeneration mechanism were studied. This material allowed a hydrogen production of 2500 NL.h-1.kgcat^{-1}, with a purity of 85% vol. corresponding to 21.3 NL.L_{fuel}^{-1} which can generate an electrical power supplied by a fuel cell of 2.5 kW. This process allows a hydrogen production on-board or in distribution sites.

[1] ITF, I. T. F. *Reducing transport ghg emissions: opportunities and costs*; 2009

[2] Carter, D.; Wing, J. *The Fuel Cell Industry Review*; 2013

[3] U.S.DOE. *Hydrogen production and delivery: summary of annual merit review of the hydrogen production and delivery program*; 2013

[4] U.S.DOE. *Hydrogen , Fuel Cells & Infrastructure Technologies Program - multiyear research, development and demonstration plan*; 2010.

[5] Resini, C.; Lucarelli, C.; Taillades-Jacquin, M.; Liew, K.-E.; Gabellini, I.; Albonetti, S.; Wails, D.; Roziere, J.; Vaccari, A.; Jones, D. *Int. J. Hydrogen Energy* **2011**, *36*, 5972–5982

[6] Gianotti, E.; Reyes-Carmona, Á.; Taillades-Jacquin, M.; Taillades, G.; Rozière, J.; Jones, D. J. *Appl. Catal. B Environ.* **2014**, *160-161*, 574–581.

[7] Gianotti, E.; Taillades-Jacquin,M ; Reyes-Carmona, A; Taillades, G.; Rozière, J.; Jones, D. *J. Appl. Catal. B Environ. 2016, 185, 233-241*

459522 Pure Hydrogen Production Via Methane Catalytic Decomposition on a Ni-Based Catalyst

Abstract:

The CO concentration in the hydrogen stream produced from methane catalytic decomposition (MCD) reaction on Ni-based catalyst has been an interesting topic. In this work, the active and stable Ni based catalysts derived from hydrotalcite-like precursor prepared by coprecipitation method were employed for the MCD reaction. To avoid oxygen impurities in the feed gas, a high purity undiluted methane and two oxygen traps (purchased from Grace Davison Discovery Sciences) were used in the reaction. The effect of the reduction temperature, reaction temperature and metal loading amount on the CO content in the exhaust from MCD reaction were studied. The results show that the reduction temperature has almost no effect on the activities of Ni-Cu-Al catalysts, but the CO formation concentration decreases with the increase of the reduction temperature. The CO concentration and methane conversion of Ni-Cu-Al catalyst both increase with the increase of reaction temperature. The results of metal loading effect on CO formation show that the CO concentration decreases with the increase of Ni loading. Besides, it shows that the oxygen all comes from the support and the incomplete reduction of metal oxide via the long term reaction test performed on 15Ni-3Cu-2Al catalyst. The CO concentration over the 15Ni-3Cu-2Al catalyst at the optimized condition (reduced at 700 ^{o}C and react at 600 ^{o}C) is below the tolerance concentration for PEMFCs (10 ppm) after the first hour of the reaction. This work proves the possibility of supply the hydrogen produced from MCD reaction directly to PEMFC.

451924 Production of Hydrogen from Various Feedstock Options Using Plasma Reforming

Abstract:

Ceramatec has developed a method for reformation of various hydrocarbons using a non-thermal plasma system to catalyze the reformation process. The plasma provides the same type of catalytic activity as a solid catalyst but since the plasma is constantly renewed it is not sensitive to sulfur or various other contaminants that adversely affect traditional solid catalysts. The reformer can operate in a partial oxidation or steam reforming mode but these modes are usually combined in an autothermal process. Any hydrocarbon that can be vaporized can be reformed with the system. The reformer can operate with air, oxygen enriched air, or pure oxygen as an oxidant. The reformer has been deployed in the field in sizes as large as 1 MMSCF per day of natural gas. The paper will discuss the technology associated with plasma reforming, the results using various feedstock options (natural gas, high sulfur diesel fuels, condensed solvents, and residuals from biomass gasification), and current programs for further development.

471282 Computational Fluid Dynamic Modeling of a Microplasma Fuel Reformer

Abstract:

Batteries are present in everyday life as the primary means to provide portable power to most of our electrical devices. Due to the low energy densities of batteries, portable fuel cell technologies have been heavily considered as a potential replacement. Fuel cells have the beneficial advantages of higher overall efficiencies, no moving parts, and low temperature operation, but exhibit the limitation of not having a readily available stream of hydrogen. Microplasma reactors have been successfully designed, fabricated, and experimented as a novel fuel reformer. However experimental measurements from the devices presented low energy efficiencies making the dynamics of the fluid transport in the novel devices an important aspect of further investigation. To learn more about the reactive microplasma environment a computational fluid dynamic (CFD) model of the microplasma reactor was created. Additionally this model will ultimately be used to inform the ideal system conditions and improve microreactor designs in future generations of the experimental devices.

The low efficiencies are likely due to the large bulk volume allowing for a large portion of the feed to pass around the microplasma. The volume of the microreactor holder is 375 μL, where the reactive microchannel is at most 0.5 μL. The CFD model was designed in COMSOL Multiphysics version 5.2. The object of study is the channel and the bulk volume above the reactor. In the model only the channel volume is reactive. The model captures both momentum and mass transport. As reactants are fed into the holder in a gas phase they pass over and through the microchannel, the material in the microchannel is reacted and then products diffuse into the bulk volume, while more reactant diffuses into the channel. Based on the inconsistency of the microplasma size throughout experiments, three model variations were developed for each experiment that accounts for the 100% of the channel being reactive as well as 5% and 50%. Each model simulates experimented results to determine accuracy of the model. Due to reactor variation the model are also parameterizable with respect to width, depth, oxide thickness, inlet flow rate, reaction rate and electric current.

To date all models have been tested with a carbon dioxide decomposition reaction; forming oxygen and carbon monoxide. The carbon dioxide reaction study was performed to analyze the microreactors, where carbon dioxide was chosen due to its ease in developing a plasma and simplicity in products. Currently the CFD model has presented reaction rates from 125 to 18,153 mol/m^3-s (2.5E-8 to 9E-8 mol/s). This work aim is to not only determine unique reaction characteristics of the microplasma environment, but also to determine key parameters to study optimization of device performance. We plan to present the results of the model, and the ideal parameters to obtain the highest energy efficiency of the microplasma fuel reformer.

462329 Syngas Production Using Steam/Red Mud Gasification of Coal

Abstract:

Coal is one of the major sources of fossil fuels used for power generation in the USA. Being one of the most polluting fuels, its use significantly contributes to environmental pollution and greenhouse gas emissions. Coal gasification is an alternative to mitigate pollutant emissions. Syngas produced by coal gasification can be used for electricity generation, chemical feedstocks, or production of synthetic natural gas. Gasification of coal using steam and catalysts provides more flexibility in the control of the hydrogen and carbon monoxide ratio, which is critical in converting syngas intro transportation fuels such as gasoline and diesel using the Fischer-Tropsch process. Herein, in this work, sub-bituminous coal was gasified at atmospheric pressure in a bubbling fluidized bed reactor at 700, 800, and 900 °C on red mud (RM) under steam. During coal gasification on RM under steam the classic water-gas shift reaction (WGSR) occurred. This reaction, which consists of reaction of carbon monoxide and water vapor to form carbon dioxide and hydrogen was promoted by the RM. Thus, the syngas was significantly enriched with hydrogen. The RM also promoted the cracking of the tars at relatively low temperatures compared to those reported in literature. Gasification of coal using RM and carbon dioxide also increased the carbon monoxide content and increased the hydrogen content through the WGSR. Addition of biomass to the coal improved the hydrogen content of the syngas because of WGSR which converted the large fraction of CO produced by the biomass to hydrogen. Additionally, the use of red mud in the gasification process prevents the formation of clinker in fluidized bed reactor.

452426 Hydrogen Production from Lignocellulosic Biomass Residues Via Gasification in Supercritical Water Catalyst Activity and Process Optimization Study

Abstract:

Abstract

Hydrogen production from biomass is an attractive alternative for clean energy generation. Supercritical water gasification (SCWG) is a promising technology for effectively conversion of wet biomass into a pressurized, clean and hydrogen rich gas [1]. Catalysis is crucial for achieving promising conversion of biomass at lower temperature, which could bringing down the cost of this process [2].

In this study, firstly, characterization and evaluation for two groups of catalysts for hydrogen production via SCWG process are carried out. For Ni-Co/Mg-Al catalysts, it was found that higher hydrogen yield could be correlated to strong acidic sites in the range of 400 -600 °C. For catalyst preparation co-precipitation is better than impregnation in terms of hydrogen yield. Also, for Ni/TiO_2 catalysts, 5 wt% Ni loading is optimum for hydrogen production, however, adding promoters such as Ru, Co, Ce, Mg or change in the active metal to Ru at 5 wt% loading did not improve the hydrogen yield.

By catalysts screening, K_2CO_3 and 20Ni-0.36Ce/Al_2O_3 were identified as the best catalysts. Then, an optimization study relating the effects of different parameters based on Taguchi experimental design was conducted, and biomass including wheat straw, canola meal, and timothy grass were used as feedstock. The order of relative importance for process parameters for hydrogen production is: temperature   catalyst loading   catalyst type   biomass type. High temperature (~650 °C), and high catalyst loading (~100%) are favorable for hydrogen production. Using different waste biomass, the average hydrogen yield was in the order of: canola meal > wheat straw > timothy grass.

Key words

Supercritical water (SCW), hydrogen, biomass, catalysis

References

[1] R.M. Navarro, M.C. Sanchez-Sanchez, M.C. Alvarez-Galvan, F.d. Valle, J.L.G. Fierro. Hydrogen production from renewable sources: biomass and photocatalytic opportunities. Energy & Environmental Science. 2 (2009) 35-54.

[2] Y. Guo, S.Z. Wang, D.H. Xu, Y.M. Gong, H.H. Ma, X.Y. Tang. Review of catalytic supercritical water gasification for hydrogen production from biomass. Renewable and Sustainable Energy Reviews. 14 (2010) 334-43.

468442 The U.S. Department of Energy's R&D Program for Carbon Use and Reuse

Abstract:

The U.S. Department of Energy Office of Fossil Energyâ€™s National Energy Technology Laboratory has an active carbon capture and storage (CCS) program. Â The goal of the Carbon Use and Reuse Focus Area is to identify and develop a suite of technologies to beneficially use CO_2 to produce useful products that can generate revenue to offset capture costs associated with CCS implementation. Identification of candidate chemistries and economically feasible approaches that utilize large amounts of CO_2 as a feedstock for high-demand product is very challenging while also ensuring that the overall process has a smaller greenhouse gas footprint that traditional pathways.

462065 Economic Production of Syngas from CO_2 and H_2O: New Routes to Synthetic Fuels

Abstract:

Economic production of syngas from CO_2 and H_2O could produce new routes to synthetic fuels. In this paper we will describe recent work on the electrolysis of CO_2 in a 5 cm^2 alkaline electrolyzer using standard Fuel Cell Technologies fuel cell hardware. Previous work had demonstrated CO_2 conversion with steady state currents of 50 mA/cm^2 or less at 3 V. In this paper we find that by optimizing the composition of the membrane and ionomer we can raise the current to 280 mA/cm^2 at 3V and 600 mA/cm^2 at 3.25-3.35 V.

Table 1 shows how the performance of the electrolyzer varies with membrane composition for several different membranes using bare silver cathode catalysts, ruthenium anode catalysts and a variety of membranes. In all cases the measurements were done by sandwiching the membrane between two pieces of carbon cloth that were coated with membrane, mounting the resultant MEA in 5 cm^2 fuel cell hardware, supplying humidified CO_2 to the cathode, applying 3 V, waiting 20 minutes to 1 hr for the current to stabilize and recording the current.

Membrane	Faradaic Efficiency	Current at 3V (mA/cm^2)	Current for carbon containing products (mA/cm^2)
Nafion 117	0%	72	0
Neosepta	34%	24	8
Acid doped CMI-7000	0.02%	35	0.007
Alkali doped AMI-7001	25%	10	2.5
SPEEK	2.5%	5	0.13
Alkali doped PVA	52%	15	7.5
Alkali doped PEI/PVA	16%	10	1.6
H_3PO_4 doped PBI	14.7%	8	1.2
Sustainion™ X1	95%	80	75
Sustainion™ X2	93%	80	72
Sustainion™ X3	83%	25	20.8

Notice that the faradaic efficiency for the conversion of CO_2 to CO varies from 0 to 95%. While the current varies from 5 to 80 mA/cm^2. Sustainion™ membranes give higher currents and faradaic efficiency than the others. These results demonstrate that the membrane composition makes a significant difference to the electrolyzer performance.

We have also determined shows how the polarization curve of the cell varies as we add different amounts of Sustainion™-X5 to the cathode layer in a cell with a Sustainion™-X4 membrane. Notice that one can obtain currents of 1.4 amps (280 mA/cm^2) at 3 volts with a Sustainion™-X4 membrane and and Sustainion™-X5 ionomer.

Steady state runs were done at 200 and 600 mA/cm^2. The cell is able to sustain 200 mA/cm^2 a 2.9 V and 600 mA/cm2 at between 3.25 and 3.35 V. By comparison, the closest competitor needed almost 7 V to obtain the same current.

These results demonstrate that one can use electrolyzers with Sustanion™ membranes to convert CO to CO_2 at industrially important rates

Acknowledgement

Parts of this work were supported by ARPA-E under contract DE-AR0000345 and by 3M. The opinion are those of the authors and may not reflect the opinions of ARPA-E.

472626 Electrochemical Carbon Dioxide Reduction As an Alternative Source of Fuels and Chemicals

Abstract:

Cost-effective industrial CO_2 recycling using renewable energy sources (ECO2R) could form the basis of an artificial carbon cycle that replaces a wide range of products that are currently derived from fossil resources. ECO2R combines CO_2, water, and electricity, and converts them into useful products using engineered metal catalysts. At full scale, this technology could eliminate our dependence on fossil resources by providing an alternative source of carbon-based compounds for fuels and commodity chemicals. However, commercial fuel and chemical production via ECO2R is challenging, because the current state of the technology is not cost-effective enough to compete with conventionally manufactured fuels and chemicals already on the market.

The key cost-drivers of ECO2R are the energy efficiency, product selectivity, and current density. Based on these metrics, we have developed a model for the cost of ECO2R-derived compounds. We will use the model to compare the cost of ECO2R derived compounds to fuels and chemicals derived from traditional sources. In doing so, we will identify the technical barriers that need to be overcome to reduce the cost, including enhanced electrocatalysis [1-3] and achieving high current densities needed for a cost-competitive process because of limited carbon dioxide solubility.

Opus 12 has also developed a prototype ECO2R reactor. The Opus 12â€™s prototype contains improved catalysts and a reactor design with high energy efficiency, high product selectivity and high current densities. We will present our technical progress in terms of these key performance metrics. Using these metrics, we can estimate the current cost of ECO2R and present a roadmap of improvements needed to further reduce costs to compete with traditional chemicals and fuels derived from fossil resources.

[1] K.P. Kuhl, E.R. Cave, D.N. Abram, T.F. Jaramillo, New insights into the electrochemical reduction of carbon dioxide on metallic copper surfaces, Energy Env. Sci., 5 (2012) 7050-7059.

[2] K.P. Kuhl, T. Hatsukade, E.R. Cave, D.N. Abram, J. Kibsgaard, T.F. Jaramillo, Electrocatalytic Conversion of Carbon Dioxide to Methane and Methanol on Transition Metal Surfaces, Journal of the American Chemical Society, 136 (2014) 14107-14113.

[3] F.S. Roberts, K.P. Kuhl, A. Nilsson, High Selectivity for Ethylene from Carbon Dioxide Reduction over Copper Nanocube Electrocatalysts, Angewandte Chemie International Edition, 54 (2015) 5179-5182.

461439 Enhanced Asphaltene Precipitation in Paraffinic Froth Treatment Using Carbon Dioxide

Abstract:

Paraffinic froth treatment is an essential step in oil sand surface mining operations for the removal of water and mineral solids from bitumen froth. In this process some of the asphaltenes present in the bitumen precipitate out and form aggregates with emulsified water droplets and dispersed mineral solids, accelerating the separation of solids and water from the bitumen and producing a clean bitumen product. In order to achieve the required asphaltene separation, a large volume of paraffinic solvent such as pentane or hexane is added to bitumen froth.

It has been reported that CO_2 dissolved in heavy oil results in asphaltene precipitation during enhanced oil recovery. Because asphaltene precipitation is considered a key step of paraffinic froth treatment, we conducted a lab-scale investigation to explore whether adding CO_2 to paraffinic froth treatment solvent could enhance asphaltene precipitation.

Froth treatment tests were conducted using both supercritical fluid CO_2 and liquid CO_2 mixed with pentane-diluted bitumen froth in an autoclave. In these tests both pentane and CO_2 concentrations were varied. The results indicated that when mixing CO_2 with n-pentane-diluted bitumen froth significant asphaltene precipitation was observed. At a pentane-to-bitumen ratio of 1.0 and lower, which is below the onset asphaltene precipitation point with solvent alone, adding CO_2 to the diluted bitumen produced significant asphaltene precipitation and resulted in a much reduced asphaltene content in the bitumen product. The combined solvent-to-bitumen ratio (CO_2 plus solvent) was much lower than that using n-pentane alone when precipitating the same amount of asphaltenes. For example, at the combined solvent-to-bitumen ratio of about 1.4, the asphaltene content in the product bitumen decreased to about 5-7 wt% from an initial 17 wt%. To reach these levels of asphaltene content using pentane alone, pentane-to-bitumen ratios of 2.2-2.5 are needed. Low water content (below 0.3 wt%) in the diluted bitumen phase was also observed. These results suggest that CO_2, when dissolved into diluted bitumen, may enhance the paraffinic froth treatment process.

450850 CO2 Conversion to Novel Solid Materials

Abstract:

In the recent years, a great effort was made for CO2 conversion by my group, leading to novel approaches. In this presentation, advances in CO2 conversion will be briefly reviewed, our important results for CO2 conversion to carbon nitrides and 3D graphene will be discussed, and the energy application of those solid materials from CO2 will be demonstrated.

466759 New Bimetallic Catalysts for CO_2 Hydrogenation to Chemicals and Fuels for Sustainable Energy Development

Abstract:

Abstract for AIChE CO2 Symp – Nov 2016

New Bimetallic Catalysts for CO_2 Hydrogenation to Chemicals and Fuels for Sustainable Energy Development

Professor Chunshan SONG*

EMS Energy Institute, Dept of Energy & Mineral Engineering, and Dept of Chemical Engineering, Pennsylvania State University, University Park, PA 16802, USA

*Email: csong@psu.edu; Tel: 814-863-4466; Fax: 814-865-3573

Capturing CO_2 and converting it into chemicals, materials, and fuels using H_2 produced with renewable energy, is an important path for sustainable development and a major challenge in 21st century. CO_2 has become a plentiful source of carbon, and there are long-term merits in using CO_2 as a feed for chemicals and materials and as a carrier for energy. The concentrated CO_2 can be used for manufacturing chemicals (lower olefins such as ethylene and propylene, methanol, and carbonates), and fuels (such as liquid transportation fuels including gasoline, diesel and jet fuels, or synthetic natural gas). Effective use of catalysis and adsorption over nano-porous materials can decouple the traditional link between energy utilization and negative environmental impacts. As an example, the hydrogenation of

CO_2 (using H_2 produced from H_2O using renewable energy) over novel bimetallic catalysts can selectively produce either C_2-$C_4^{=}$ olefins (and C5+ liquid hydrocarbons) using Fe-Co catalysts, or methanol over Pd-Cu catalysts. Nano-structured catalysts based on hollow zeolite-encapsulated Ni nanoparticles can create a highly stable and carbon-resistant catalysts for co-conversion of CO_2 and CH_4 in bio-gas to syngas for chemicals and ultra-clean fuels.

468853 Combined Magnesium OXIDE/Water Gas Shift-Based CO2 Capture Process

Abstract:

COMBINED MAGNESIUM OXIDE/WATER GAS SHIFT-BASED CO2 CAPTURE PROCESS

Andrew Lucero, Jiajia Meng, Shen Zhao, Santosh Gangwal

Southern Research, 5201 International Drive, Durham, NC 27712

alucero@southernresearch.org; (919) 282-1054

Southern Research is developing a combined magnesium oxide (MgO) CO2 sorbent/water gas shift (WGS)-catalyst-based CO2 capture process for integrated gasification combined cycle (IGCC) power plants. A heat-exchanger reactor with a highly efficient heat management capability is being developed for simultaneous CO2 capture and WGS to convert CO to CO2 and H2 and capture more than 90 % or more of the carbon from coal gasifier syngas. Simultaneous heat management and reaction allows the maintenance of thermodynamically favorable reaction temperatures for both steps. Also, by simultaneously converting the CO to CO2 over the shift catalyst, the overall reaction thermodynamics is shifted allowing greater carbon capture by the MgO sorbent than if the two steps were carried out separately. The ultimate project goal is to capture > 90 % CO2 and to reduce the cost of electricity by 30% over IGCC plants employing conventional methods of CO2 capture.

Extensive thermodynamic analyses from multiple sources of thermodynamic data have been completed for simultaneous CO2 capture and water-gas shift reactions. Based on these analyses, optimum temperature and CO2 partial pressure windows have been established for reactor design. A thorough analysis of the literature has been conducted for preparation of MgO sorbents with excellent multi-cycle durability and high CO2 capture capacity. An appropriate commercial water-gas shift catalyst has been obtained from a reputable catalyst manufacturer. Several MgO-based sorbents with and without promoters have been synthesized and have been tested using a thermo-gravimetric analyzer (TGA) at various temperatures and CO2 partial pressures determined from the thermodynamic analysis. The best performing sorbents have shown CO2 capture capacities greater than 20 wt %.

Present sorbent development/synthesis efforts are concentrating on (i) achieving excellent multi-cycle durability for these high capacity sorbents over 100s of cycles and (ii) mechanical forming of the sorbent in appropriate form for use in the heat-exchange reactor along with the WGS catalyst. A laboratory scale CO2 capture/ WGS process skid has been designed and constructed to test sorbents under relevant gasifier conditions in a ½ inch diameter, 18 inch long reactor. Following sorbent development and completion of laboratory scale experiments, the sorbent and process will be scaled up to bench-scale. Simultaneous heat management, CO2 capture and WGS will be demonstrated with high capacity and durability over 1000 cycles. A techno-economic evaluation will be conducted to demonstrate the potential of the process to achieve the required reduction in cost of electricity.

461222 Oxidative Dehydrogenation of Ethane and Propane By CO_2

Abstract:

A series of vanadium catalysts (V loading in the 0.7 ~ 16 V/nm^2 range) supported on silica have been prepared by incipient wetness impregnation method. The activity of ODH was studied in ethane and propane oxidative dehydrogenation in the presence of O_2, CO_2 and their mixture from 350 to 600°C and at atmospheric pressure. The products of oxidative dehydrogenation of paraffins were corresponding olefins, CO and CO_2, in the propane reaction a small part was cracked to ethylene and methane.

The vanadia silica catalysts at sub-monolayer VO_x coverage were found to be active and selective in propane ODH by O_2 and CO_2, while VO_x multilayers and nano-sized clusters showed lower activity and selectivity. Moreover, due to its acidic character of silica, it showed undesirable carbon deposition (coking) in propane ODH by CO_2 in a short period of reaction time and as a result catalyst deactivated quickly. Around 20 % weight loss observed for the used catalyst after 30 hours propane ODH by CO_2 reaction under conditions of TGA in air. On the other hand, the same catalyst employed to propane ODH by O_2, almost no carbon deposition was observed. Our experimental observations are in agreement with recent theoretical calculations indicating that in this catalytic system, carbon deposition is thermodynamically favorable, unless the surface character is modified by basic additives, such as Na^+. Moreover, basic additives were recently reported to enhance the monolayer coverage of VO_x species on silica which also improved their activity and selectivity in propane ODH by CO_2.

The catalytic activity of these species may be optimized by varying the surface coverage of VO_x species and the nature of the support and promoters.

451420 Conversion of Coal and Biomass to Liquid Hydrocarbons Using Gasification and a Hybrid Fischer Tropsch Catalyst

Abstract:

Ceramatec is leading a team to develop a two (2) BPD (barrel per day) Fischer Tropsch system to produce jet fuel. Funding is provided by the United States Air Force, through the US Department of Energy - National Energy Technology Laboratory. The feedstock for the system will be pulverized coal with some biomass to meet carbon emission limitations. The coal - biomass will be gasified and the resultant synthesis gas compressed to pressures suitable for Fischer Tropsch (FT). The synthesis gas stream will be cleaned using a sulfur removal system provided by IntraMicron, a company located in Auburn, AL. The clean synthesis gas will be fed to a 2 BPD modular FT reactor designed and assembled by Ceramatec.

Ceramatec is currently operating a small GTL facility at its Salt Lake City laboratory and portions of this facility will be used for this program also. The sulfur removal system and the 2 BPD FT reactor will be tested at this existing facility prior to shipment to the gasifier location for testing with the coal - biomass derived synthesis gas. A hybrid FT catalyst capable of terminating carbon chain growth in the liquids range will be used in the FT reactor.

The objectives and challenges of the 2 BPD CTL (coal-to-liquids) project will be described in the presentation. The presentation will also describe the capabilities of the team to address the issues associated with development of that integrated 2 BPD CTL system. The test data generated to date will be presented to demonstrate how the issues are being addressed. This data will include the carbon distribution on various test runs, catalyst performance, and synthesis gas utilization. The design for the integrated CTL facility will be described and individual components highlighted. Other relevant technical data from associated projects will be presented.

460840

The Effects of Carbon Dioxide on the Adsorption Capacity of Methane in Shale

Abstract:

- Abstract Text

The effects of carbon dioxide on the adsorption capacity of methane in shale

Bing Cao, Lu Wang, Lu Xia, Qingchun Yu

With population increases and economic development, energy consumption has increased worldwide. Consequently, balancing energy supply and demand is a priority. Shale gas is a clean and efficient non-conventional energy source that has attracted increasing attention in many countries. Generally, shale reservoirs contain carbon dioxide (CO_2). Under reservoir conditions, shale has a larger adsorption capacity for CO_2 than for methane (CH_4), especially in primary pores. Recent research showed that five CO_2 molecules could displace one CH_4 molecule. Therefore, the influence of the CO_2 content on the adsorption of CH_4 in shale is not negligible. There have been a number of national and international studies conducted on the adsorption of CH_4 and CO_2 mixtures by shale to evaluate the effect of CO_2 on CH_4 adsorption.In the present study, two shale samples were obtained from depths of about 838.5 m and 1035.4 m. To reduce the equilibration time of the sorption, the shale samples were pulverized to powders and screened through an 80 mesh sieve. Both of the samples were analyzed by scanning electron microscopy, and for their vitrinite reflectance and total organic carbon content. The experiments were conducted with a gas-solid two-phase adsorption instrument, which included a gas pressurization system, gas adsorption device, data acquisition system, vacuum pumping system, and gas composition analysis system, all designed by our research group. The gases injected into the pressurization system and used for adsorption were CH_4, a binary mixture of 95.06 % CH_4 and 4.95 % CO_2, and another binary mixture of 80.01 % CH_4 and 19.99 % CO_2. The gas adsorption measurements were performed at 303.15 K, 313.15 K, or 323.15 K and at a pressure between 0 and 10 MPa. For both CH_4 and the binary mixtures of CH_4 and CO_2, the absorption capacity decreased with increasing temperature but increased with increasing pressure. In addition, absorption of the CH_4/CO_2 mixture on the shale surface was affected by the total organic carbon content. As the total organic carbon content increased, the absorption was enhanced by the increased interfacial area between the gas and shale surface. The difference between the results for pure CH_4 and the mixture containing a small amount of CO_2 was not pronounced when the experiments were conducted at the same temperature and low pressure.Therefore, in an actual reservoir, the presence of trace amounts of CO_2 will likely have little effect on the adsorption of CH_4. However, the difference increased if either the amount or pressure of CO_2 was increased. Although the CO_2 content is low, the adsorption capacity for CO_2 is stronger than that for CH_4. This could occur because CO_2 molecules may be better accommodated in the pore structure of shale than CH_4 molecules. These results indicate that CO_2-enhanced shale gas recovery could be feasible for shales in the Qaidam Basin (China). Further tests and analysis on the adsorption behavior of each gas will be conducted at higher CO2 fraction. Results from this study may help in understanding the CH4 adsorption behavior and the effect of CO2 at low fraction (<10%), which is representative for realistic reservoir conditions.

468490 Unconventional Oil and Natural Gas Science & Technology Advancement

Abstract:

Unconventional oil and natural gas (UOG) resources, which include resources contained in shale and other low-permeability geological formations, are transforming the energy industry in the United States and around the world. Rapid growth in the availability of UOG resources is an opportunity, but recovery efficiencies are commonly low in these reservoirs. Current development approaches are not yet achieving an optimal balance of high resource utilization and low environmental impact. The United States Department of Energy, through its National Energy Technology Laboratory (NETL), has an active UOG program. This presentation will focus on critical knowledge gaps in our understanding of the operational and environmental optimization of UOG operations, and the technological improvement and operational development of subsurface hydraulic fracture geometry and optimal completion/stimulation strategies for the most efficient recovery of UOG resources.

454817 Well Integrity Inspection in Unconventional Gas Wells

Abstract:

The inspection of multi-casing/multi-cement-annuli in hydrocarbon producing wells is of paramount importance for the protection of the environment and to guarantee long term integrity of production wells. However, state-of-the-art imaging technologies used for monitoring the integrity of hydrocarbon-producing well bores are typically limited to inspection of the innermost metal casing and its surrounding cement bond interface. The scope of this work is the development of a multi-casing/multi-annuli inspection system for providing information about the flaw structure and topology of unconventional gas wells. The proposed well integrity inspection system utilizes high energy gamma-rays for deep penetration into metal casings and cement structures. The high-energy imaging modality is complemented by conventional imaging techniques such as ultrasound or electromagnetics.

Monte-Carlo N-Particle (MCNP) simulations and laboratory prototype testing suggest that it is possible to detect air voids and water defects well past the innermost casing. Numerical simulations clearly distinguish air voids with diameters less than ⅛" for several inches into the well bore structure. Preliminary data from our fist gamma-ray prototype clearly show evidence of a ¼" diameter air void in the first metal casing and a ½" diameter air void in the second metal casing. Both metal casings are separated by a cement annulus with a thickness of 1". Design improvements of this preliminary gamma-ray prototype are ongoing and further enhancement of the detection limit is expected. The high energy gamma-ray modality can be used to complement other more conventional imaging modalities such as ultrasound or electromagnetics. Current development of an electromagnetic-based imaging tool shows promising performance in a laboratory environment. An array of electromagnetic transducer coils is used to measure the material loss in the first and second casing as well as to detect the eccentricity between the two casing strings. However, the electromagnetic imaging tool cannot detect defects in non-conductive materials such as the cement annulus structure. By combining imaging modalities (e.g., gamma-ray and electromagnetics) into a single inspection system and by fusion of data from each sensing modality, we hope to further improve the detection limit of well integrity monitoring systems. Current efforts investigate benefits of data fusion between high energy modalities and conventional imaging techniques. It is expected that independently acquired data will provide enhanced detection limits and better accuracy compared to the sum of individually acquired data. A multi-modality inspection system with imaging capabilities for metal and concrete defects well past the first casing should be greatly beneficial for well integrity inspection of multi-casing wells at intermediate-to-surface depths along major ground water zones. We will present a tradeoff analysis of performance parameters such as minimum defect size, sector angle resolution, depth resolution, penetration depths, and logging speed. Benefits and drawbacks of the proposed technology will be discussed.

470387 4D Synchrotron X-Ray Imaging to Understand Porosity Development in Shales during Exposure to Hydraulic Fracturing Fluid

Abstract:

Unconventional or low permeability shale reservoirs have emerged as the most important source of petroleum resources in the United States and represent a two-fold decrease in greenhouse gas emissions compared to coal. While individual plays differ significantly in lithology, P-T conditions and depositional settings, unconventional reservoirs are united by necessity of hydraulic fracturing for economically viable production. The proliferation of modern fracturing techniques, specifically horizontal drilling and multi-stage fracturing, has generated a significant increase in oil and gas production rates in the last decade alone. Despite recent progress, further optimization of hydraulic fracturing operations presents substantial technical, economic, and environmental challenges. These include inefficient recovery, wastewater production and disposal, contaminant and greenhouse gas pollution, and induced seismicity.

A relatively unexplored facet of hydraulic fracturing operations is the fluid-rock interface, where hydraulic fracturing fluid (HFF) contacts the shale matrix along pre-existing and stimulated faults and fractures. Widely used, water-based fracturing fluids (slickwater) contain additives not found in shale, including oxidants and acid. Consequently, this creates an environment significantly out of equilibrium with the shale minerals and trapped hydrocarbons. Injection and soaking during hydraulic fracturing induces a host of fluid-rock interactions, most notably the dissolution of carbonate and sulfide minerals, which results in enhanced or "secondary" porosity networks, as well as mineral precipitation. The competition between these mechanisms determines how HFF affects reactive surface area and permeability of the shale matrix. The resultant microstructural and chemical changes may also affect the formation of capillary barriers for trapped hydrocarbons and facilitate the release of heavy metal contaminants from organic matter. At present, few of these processes have been investigated, specifically how they impact production and the composition produced waters. Developing a mechanistic understanding of the microstructure and chemistry of the shale-HFF interface enables the design of new methodologies and fracturing fluid compositions to improve estimated ultimate recovery (EUR) while minimizing the potential for contaminant release from produced water.

Synchrotron-based X-ray imaging techniques provide unique advantages for studying dynamic processes in space (3D) plus time (4D). With a high brilliance synchrotron light source, tomographic datasets can be collected in minutes to seconds (depending on the instrument) and non-destructively capture time-dependent phenomenon. In addition, 2D projection images can also be collected at higher temporal frequency to visualize more rapid processes. By employing various configurations, optics can be applied in the visible and X-ray regimes to magnify the object to measure microscale to nanoscale features. X-ray microtomography provides submicron voxel size which can resolve the secondary pore network which is forming while X-ray nanotomography can further magnify regions of the sample and examine the nanoporous structure. This flexibility in different synchrotron-based microscopes provides a unique capability to quickly image and reconstruct samples at various length scales relevant to shales exposed to HFF.

To assess the effects of HFF on shale reservoir rock, imaging experiments were carried out at two synchrotron facilities, Pohang Accelerator Laboratory (PAL) and Stanford Synchrotron Radiation Lightsource (SSRL). Shale samples were collected from both outcrops (Green River, Marcellus) and vertical wells (Barnett, Eagle Ford) and span a range of lithologies from siliceous to calcareous to organic-rich. Sample were imaged before, during, or after exposure to HFF to characterize the initial 3D structure. The subsequent microstructural evolution was tracked in time using time resolved 3D imaging. By capturing time series, the 3D reconstructions reveal how the secondary porosity networks advance into the shale matrix. By testing shales of different lithologies we have obtained insights into the mineralogic controls on secondary pore network development and the morphologies obtained at the shale-HFF interface and the ultimate composition of produced water from different facies. Utilizing the information gained from the beamline imaging experiments provides the information about thickness of reacted "skin" around fractures need to develop new methodologies and fracturing fluid compositions aimed to enhance EUR and minimize environmental impact.

472953 Changes in Bubble Point of Hydrocarbons in Shales: Effect of Confinement and Presence of Kerogen

Abstract:

The exploration and production from shale plays have revolutionized oil and gas production in US. The fluid properties in low permeability, low porosity shales are, however different from the conventional reservoirs. The kerogen is a organic matter found in shale rocks. It is considered as precursor to oil and gas found in shales and is a key component of the shale rock. The pore sizes in organic and inorganic matrix in shales are found to be in the range of nanometers. Thermodynamics of the oil and gas in intergranular porosity and in kerogen-related porosity are topics of significant research. The saturation pressures of hydrocarbons changes due to confinement and presence of kerogen in shales. Molecular Dynamics Simulations of kerogen-oil system and experiments related to swelling of kerogen with oils show the interaction of kerogen with oil.

In current work, kerogen is isolated from a shale sample by a series of acid treatments. The shale rock sample was collected from one of the prolific US shale play. The isolated kerogen was then swelled with octane-decane (60-40%) mixture. Swelling of kerogen is measured using Differential Scanning Calorimetry (DSC) and the results were compared with swelling ratios reported in published data. DSC is then used to measure the changes in bubble point of same octane-decane mixture in kerogen, at room as well as reservoir temperatures. The work, thus accounts for the effect of confinement of oil and kerogen-oil interactions on pressure-volume-temperature (PVT) properties of oil. The research shows that the bubble point pressures for oil at given temperatures are suppressed in the nano pores and in the vicinity of kerogen. The changes in saturation pressure of oils in shales effect properties like production gas-oil ratio and volumes and rates of recovery of petroleum reserves.

458746 Advanced Supercritical Water-Based Process Concepts for Treatment and Beneficial Reuse of Brine Generated By Oil/Gas Production

Abstract:

Both U.S. energy and economic security rely upon continued development of U.S. unconventional shale plays, which require access to suitably clean water resources. Produced water (brine) is the nation's largest industrial waste stream, with approximately 22 billion barrels generated annually by the U.S. oil and gas sector. The recent surge and continued development of unconventional U.S. shale plays will result in greater volumes of produced water. In addition, an additional brine waste stream will be created by future sequestration of CO_2 emissions generated by fossil-based power plants into saline aquifers. Utilizing this brine in development of unconventional shale resources represents a beneficial reuse of this waste stream. This approach can not only lower stress on local watersheds but can also address public/local government concerns regarding long-term ground water contamination potential and seismic activity associated with Class II salt water disposal wells (SWDs).

However, direct reuse of brines in shale development activities currently is limited due to constituents found in this waste stream. Specifically, the high levels of dissolved solids found in the brine may cause scaling within the shale or within production casings, reducing well productivity. Current brine treatment technologies including membrane- and thermal-based processes are ineffective in treating brine-containing concentrations of dissolved solids greater than 80,000 ppm due to fouling or cost/sizing, respectively.

To address these limitations, Ohio University (OHIO) with funding from the U.S. Department of Energy's Research Partnership to Secure Energy for America (RPSEA) Project 11122-60 and the National Energy Technology Laboratory (NETL) Crosscutting Research Program Project DE-FE0026315, has been developing an advanced supercritical water (SCW)-based process for treatment and beneficial reuse of brine waste streams. This SCW process offers an advantageous media for brine treatment, as its lower fluid density and decreased hydrogen bond strength provides a means to simultaneously remove dissolved solids and hydrocarbons.

Previous SCW-based brine treatment systems have been plagued by internal scaling, resulting in inefficient heat transfer, plugging, and process downtime. To address this issue OHIO has been developing both externally- and internally-heated SCW reactor design concepts, which utilize advantageous fluid dynamics and electrically driven Joule-heating mechanisms, respectively. OHIO's new SCW reactor designs offer the potential to provide a field deployable brine treatment process and a resultant product which may be reused in shale development or other beneficial reuse applications, thereby supporting goals of NETL's Strategic Center for Oil and Natural Gas.

To evaluate process potential, OHIO has been conducting both experimental investigations using prototype SCW reactors and process simulations/techno-economic assessments using Aspen Plus^{TM}. Removal of dissolved solids has been investigated at temperatures and pressures ranging from 377-520 °C and 23-32 MPa, respectively, demonstrating the ability to recover greater than 99 percent of salts from brines containing dissolved solid concentrations greater than 280,000 ppm. In addition, minimal to no internal reactor scaling has been observed. Techno-economic summaries have identified important brine treatment cost sensitivity parameters and heat recovery schemes to lower brine treatment costs. This presentation will review experimental and techno-economic study results from both U.S. DOE supported projects.

465402 Adsorption, Desorption and Flow Phenomena of Methane-Ethane Binary Gas Mixtures in Shale Powder and Whole Core Samples

Abstract:

Methane (CH_4) is the largest component of the natural gas produced from shale rocks. Although methane accounts for 87-96 mol% of the gas from these rocks, other components including hydrocarbons such as ethane (C_2H_6) and propane (C_3H_8), nitrogen (N_2) and helium (He) are also found in the gas produced. Methane and the various hydrocarbons are stored in the adsorbed state in the micro and mesopores of the shale rocks, and as free gas in the fracture networks. Though convective and diffusive transport accounts for the short-term behavior in gas production, desorption is thought to dominate the long-term dynamics of shale-gas generation.

The key objective of this study is the investigation of adsorption/desorption and diffusive and convective transport phenomena of methane/hydrocarbon mixtures in shale-gas rocks. In this talk, in particular, we focus our attention on the binary CH_4/C_2H_6mixture, since ethane is typically the second largest component in shale gas, and is thought to compete for the same adsorption sites in shale-gas rocks as methane. We study the adsorption/desorption behavior of this mixture (and its individual components) in ground and whole shale rock samples using thermogravimetric analysis (TGA). The sorption isotherms generated are important to predict the gas storage capacity of the shale samples, while the study of adsorption/desorption dynamics/kinetics help us understand the role of desorption during the later times of gas production. The combination of experiments with whole and ground samples allow us to isolate desorption kinetics from diffusive and convective mass transfer. This, in turn, facilitates our modeling and interpretation of the experimental observations.

The extended Langmuir model and the more sophisticated multicomponent potential theory of adsorption (MPTA) approach are used to model the experimental observations, and together, they provide for improved predictive

capabilities of the shale gas equilibrium and dynamic characteristics. In this work, we also study the dynamics of shale-gas production from full-diameter shale-gas cores, under realistic field-scale pressure, temperature, and confining-pressure conditions. Depletion experiments from such cores storing CH_4/C_2H_6 mixtures at various compositions (simulating a model shale-gas) are carried-out, during which the pressure of the core, the flow of gas, and its composition are all continuously monitored and measured. Our experimental observations clearly indicate a lag in ethane production relative to methane due to the preferential sorption of ethane on the shale samples. These experiments, and related analysis and modeling, nicely compliment the studies with the ground and the cube shale samples, and help one to better understand the interplay between adsorption/desorption and transport phenomena that take place during shale-gas production from these rocks.

464351 Phase Behavior of Methane Hydrates in Confined Media

Abstract:

Hydrate phase equilibria for the binary methane and water mixture were measured by high-pressure differential scanning calorimetry in mesoporous and macroporous silica that had controlled pore size ranging from 8 nm to 200 nm. An oscillating dynamic method was used to form hydrates and determine the decomposition conditions of methane hydrates. Significant shifts of dissociation temperatures were observed and fitted by modified Gibbs-Thomson equation. The experimental data up to 50 MPa indicated the interfacial energy in the confined media showed a pressure dependency. Enthalpy of hydrate fusion in confined media was experimentally measured and calculated by Clapeyron equation. Both methods indicated that the enthalpy of fusion was independent of pore size. The effects of interfacial energy on phase equilibria were investigated quantitatively by grafting of the silica surface. The dissociation temperatures of methane hydrates in hydrophobic and hydrophilic silica were significantly different. Methane hydrates in strong hydrophobic porous silica tended to behave as bulk methane hydrates regarding the equilibrium temperature. The measured phase equilibria data, interfacial energy, and enthalpy were considered as relevant parameters for different gas hydrate simulation models.

462509 Global Optimization of Constrained Grey-Box Models for Well Injection and Production

Abstract:

Crude oil development and production using secondary enhanced oil recovery techniques increase oil production by 20 to 40 percent by injecting water to oil fields to displace and drive oil to production wellbore [1-2]. Secondary recovery is a water-intensive method and the future availability of water resources is greatly threatened due poor water management in oil production fields (roughly 210 million barrels of water produced for every 75 million barrels of oil) [3]. This creates a huge demand to design and operate, environmentally sustainable and optimal well-control methods in waterflooding of oil reservoirs. However, waterflooding well-control technology in well injection and production greatly suffers from the curse of dimensionality. Thus, the objective of this paper is to introduce a sustainable innovative water-control technology based on a new optimization framework that significantly reduces the computational effort for waterflooding well-control optimization problems by (a) reducing the dimensionality of the optimization formulation, and (b) using a constrained grey-box optimization algorithm, which couples deterministic global optimization with surrogate modeling.

The first step of this workflow is based on parameterization of the well-control variable domain through a set of functional relationships, which is denoted as Functional Well-Control Method (FCM). The functional forms can range from Polynomial to Exponential or Hybrid Control Methods (PCM, ECM, HCM). Through this approach, we transform the optimization search space from the traditional pressure-based or rate-based control to a reduced space formed by the coefficients of the selected functional method. These new formulations are then optimized by the ARGONAUT algorithm [4-5], which has been developed for constrained derivative-free optimization problems. This optimizer is comprised of several mixed-integer and/or nonlinear optimization sub-problems for (a) sampling selection, (b) surrogate model identification and parameter estimation and, (c) global optimization of the formulated constrained surrogate formulations. The resulted non-linear constrained optimization problems are solved to global optimality using

deterministic global optimization solver ANTIGONE [6-9], which is a state-of-the-art global solver for mixed-integer nonlinear optimization, used in a wide range of applications.

We test the efficiency of the entire framework, with and without constraints, on a realistic three-dimensional model (the UNSIM-I-D benchmark) [10]. Our results demonstrate significant computational savings due to the coupling of ARGONAUT [4-5] and the FCM formulation. The FCM leads to a substantial reduction in the number of control parameters as we seek the optimal function coefficients to describe the control trajectories as opposed to directly searching for the optimal control values at each time interval. In addition, we compare our results with other gradient-free and gradient-based algorithms which have been traditionally used in the literature (such as NOMAD [11-12] and EGO [13-14]) and we demonstrate that our framework leads to improved solutions with higher consistency and with reduced sample-calls to the reservoir simulation.

References:

1. CA Floudas, AM Niziolek, O Onel, LR Matthews, Multi-Scale Systems Engineering for Energy and the Environment: Challenges and Opportunities. AIChE Journal, 62(3), 2016, 602-623.
2. United States Department of Energy, Office of Fossil Energy: Enhanced Oil Recovery, (n.d.). http://energy.gov/fe/science-innovation/oil-gas-research/enhanced-oil-recovery.
3. B Bailey, M Crabtree, J Tyrie, J Elphick, F Kuchuk, C Romano, *et al.*, Water control, Oilfield Review,12, 2000, 30-51.
4. F Boukouvala, MMF Hasan, CA Floudas, Global Optimization of General Constrained Grey-Box Models: New Method and Its Application to Constrained PDEs for Pressure Swing Adsorption, Journal of Global Optimization, 2016, DOI: 10.1007/s10898-015-0376-2.
5. F Boukouvala, CA Floudas, ARGONAUT: Algorithms for Global Optimization of Constrained Grey-Box Computational Problems, Optimization Letters, 2016, DOI: 10.1007/s11590-016-1028-2.
6. R Misener, CA Floudas, GloMIQO: Global Mixed-Integer Quadratic Optimizer. Journal of Global Optimization, 57(1), 2013, 3-50.
7. R Misener, CA Floudas, A Framework for Globally Optimizing Mixed-Integer Signomial Programs, Journal of Optimization Theory and Applications, 161(3), 2014, 905-932.
8. R Misener, CA Floudas, ANTIGONE: Algorithms for coNTinuous Integer Global Optimization of Nonlinear Equations, Journal of Global Optimization, 59(2–3), 2014, 503–526.
9. F Boukouvala, R Misener, CA Floudas, Global Optimization Advances in Mixed-Integer Nonlinear Programming, MINLP, and Constrained Derivative-Free Optimization, CDFO, European Journal of Operational Research, 252(3), 2016, 701–727.
10. AT Gaspar, GD Avansi, AA dos Santos, JC von Hohendorff Filho, DJ Schiozer. UNISIM-ID: Benchmark Studies for Oil Field Development and Production Strategy Selection. International Journal of Modeling and Simulation for the Petroleum Industry, 9(1), 2015, 47-55.
11. MA Abramson, C Audet, G Couture, JE Dennis, C Tribes, The NOMAD Project (2009). https://www.gerad.ca/nomad 52.
12. S Le Digabel: Algorithm 909: NOMAD: Nonlinear Optimization with the MADS Algorithm. ACM Trans. Math. Softw. (TOMS) 37(4), 2011, 1–15.
13. DR Jones, M Schonlau and WJ Welch: Efficient Global Optimization of Expensive Black-Box Functions, Journal of Global Optimization, 13, 1998, 455-492.
14. MJ Sasena, P Papalambros, P Goovaerts, Exploration of Metamodeling Sampling Criteria for Constrained Global Optimization, Engineering Optimization, 34(3), 2002, 263-78.

457852 An Overview of the U.S. Doe Carbon Storage Program's Approach to Developing Safe, Effective Commercial Carbon Storage Technologies

Abstract:

The Carbon Storage Program implemented by the U.S. Department of Energy's (DOE) Office of Fossil Energy and managed by the National Energy Technology Laboratory (NETL) is helping to develop technologies that safely and permanently store carbon dioxide (CO_2) without adversely impacting natural resources or hindering economic growth. Specifically, the Carbon Storage program is developing and advancing carbon capture and storage (CCS) technologies both onshore and offshore that will significantly improve the effectiveness of the technology, reduce the cost of implementation, and be ready for widespread commercial deployment. Since 1997, DOE's Carbon Storage program has significantly expanded the CCS knowledge base in selected technology areas through a diverse portfolio of applied research projects and field demonstrations. The Storage program is comprised of three primary technology components: (1) Advanced Storage R&D, (2) Storage Infrastructure, and (3) Sub-Disciplinary R&D. These areas work together to address significant technical challenges in order to meet program goals that support the scale-up and widespread deployment of CCS.

The Carbon Storage Program pursues the advancement of fossil fuel power-generation technologies by supporting the development of technologies which follow a sequential progression of scales (stages) toward making the technology available for commercial deployment. The scale of a technology is the size of the system relative to the final scale of the application, which in this case is a full-scale commercial power production or industrial facility. As RD&D progresses, the scale of the tests increases incrementally from lab/bench scale, to pilot scale, to pre-commercial scale, and to full-commercial scale. The acceptable degree of system integration at each scale is based on the scope of the technology under development within a particular research effort. As the test scale increases, the duration and cost of the projects increase, as does the probability of technical success. In order to assess the progress towards commercialization of a specific technology, a Technology Readiness Level (TRL) concept has been adopted by the Carbon Storage Program to help guide and evaluate the RD&D process. TRLs provide an assessment of the technology development progress on the path to meet final performance specifications. The TRL for a technology is established based upon the scale, degree of system integration, and test environment in which the technology has been successfully demonstrated.

CCS and other clean coal technologies are expected to play a critical role in mitigating CO_2 emissions while supporting energy security throughout the world. DOE's Carbon Storage program has positioned the United States on a path toward ensuring that the enabling technologies will be available to address the demands of new regulations and impact CCS projects in the 2020–2030 timeframe. Continued U.S. leadership in technology development and future deployment is important to the cultivation of economic rewards and new business opportunities, both domestically and abroad.

456707 Impacts of Realistic and Varying Rates of CO_2 Injection on *Safe* CO_2 Storage in the Bunter Sandstone of the Southern North Sea

Abstract:

Carbon capture and storage (CCS) has been recognized as a key technology to reduce CO_2 emissions and mitigate global climate change. Recently, CCS was included in several countries' emissions reduction target as part of the United Nations Framework Convention on Climate Change Paris agreement. In the UK, CCS will enable us to meet an 80% target reduction in CO_2 emissions from 1990 levels by 2050 as set by the Climate Change Act 2008.

In order to maximize the potential for CCS in any region, it is vital to develop a thorough understanding of a country's potential storage reservoirs and their dynamic behavior in response to CO_2 injection. This is particularly key as the pore volume of a reservoir does not equate in a simple way to storage capacity. This is because pressure constraints limit the rates of injection. This in turn has complex dependencies on reservoir geometry, and nature of connectivity and the bounding seals.

Carbon storage in the UK is recognized as an important transition technology for mitigating emissions of fossil fuel power plants while the capacity factor of renewable energy sources increase over time. For that reason, we investigate scenarios of gradual CCS deployment, with plants starting capture at different times and eventually shutting down progressively. Accordingly CO_2 capture rates and injection into storage reservoirs vary over time.

The Bunter Sandstone Saline Aquifer in the UK Southern North Sea is believed to provide one of the largest storage potentials for the UK. In this work we have simulated dynamic injection of CO_2 into the Bunter Sandstone – using a geological model and the commercial reservoir simulator Eclipse. The study assumes that the field has closed boundaries and contains 12 injection sites. We do not consider reservoir management techniques such as brine production. Limiting the bottomhole pressure at each injection site to 75% of the lithostatic pressure imposes a conservative constraint minimizing risk to fracturing of the sealing caprocks.

Four different geographical and market-based scenarios of the UK's CCS deployment and CO_2 capture rates are fed to a pipeline network towards the Bunter Sandstone field. We have found that there may be a significant start up time delay before CO_2 injection can be achieved at the target rates provided by the plants. This is caused by limited early-time CO_2 injectivity which evolves as more CO_2 is injected into the reservoir. The effect is sensitive to reservoir permeability, depth of the injection site and the bottomhole pressure limit at the injection point. Our results show that limited CO_2 injectivity may persist for the first few years of injection and play a significant role in planning for large scale deployment of CO_2 storage.

The key outcome of this work is demonstrating the ability to use reservoir simulation to reflect prospective consequences and responses to realistic CCS deployment and reaction to CO_2 supply. Hence, the reservoir generates a feedback to the CO_2 transport network system - controlling the injection rates delivered to the storage site - the capture system - feeding into the network - and the market demand for fossil fuel products that in turn controls the CO_2 supply. Understanding this link between CO_2 supply and reservoir behavior allows us to inform decision makers on how best to develop CO_2 storage in line with capture and transport system deployment.

461703 Geologic CO_2 Storage Leakage Detection with Statistical Analysis of Controlled-Release Field Data

Abstract:

For geologic CCUS to be accepted, leak detection capability must be guaranteed. Furthermore, it is ideal for leak detection algorithms be very practical and applicable at any geologic CCUS site. The primary objective of this analysis is to develop and test algorithms for effective detection of leakage from geologic carbon capture and sequestration (CCUS) sites. Other leak detection techniques such as monitoring eddy flux and atmospheric modeling have proven effective at detecting simulated leaks CCUS sites. These methods require knowledge of turbulent flow patterns and complex boundary layer models. Monitoring eddy flux is difficult due to the significant impacts that variable terrain, buildings, seasonal cycles, diurnal cycles, and biological processes have on the flux measurement. It is evident that we need a more straightforward, probabilistic method to leak detection and identification that could be employed at any monitoring site. The techniques developed in this study have filled that need.

As part of the monitoring, verification, and accounting (MVA) efforts of the Southwest Regional Partnership, this study developed and tested a leak detection methodology using single and multiple greenhouse gas detection towers on the University of Utah campus. Controlled releases of CO2 and CH4 were used to simulate various point sources by changing duration, location, and flow rate of the point source emissions. High frequency background data (on the order of 10 Hz) and the controlled release experiment data were collected. Point source emission concentration thresholds for both CO2 and CH4 were determined by probabilistic analysis of the data. Data was also used to construct a visual "probability density rose" (PDR) diagram. A PDR is a polar diagram that represents the probability of a leak coming from a certain direction with respect to the detection tower. PDR's were constructed using CO2 and CH4 concentration data and meteorological data collected at the detection tower over a specified time span, which was typically at least 24 hours to account for varying wind directions throughout the day. The wind direction was split into a designated bin size, which for this study was 5°. The number of measurements above the concentration threshold of a leak for each bin was divided by the total number of measurements in each bin to give a "probability" of the concentration exceeding the leakage concentration threshold for each bin.

The resulting statistical analysis (both descriptive and circular statistical analyses) on the data collected indicated that there were significant differences in wind direction, wind speed and background concentration data throughout the day as well as from day to day. This confirmed the importance of collecting long periods of background data for successful leak detection methodology development. Furthermore, in most instances the direction from which readings representative of a point source emission were significantly different than the dominant wind direction. When compared to traditional wind rose diagrams and pollutant concentration rose diagrams, PDR diagrams were shown to be far more useful in determining the direction from which leakage signals were originating from. This was attributed to the probabilistic nature of the diagram and that by dividing the number of measurements above the concentration threshold by the total number of measurements in each bin, the diagram has accounted for the likelihood of readings coming from that direction. It was concluded that PDR diagrams were not only effective visual tools, but also effective at determining the leakage source location with respect to the detection tower.

Ongoing and planned future work includes conducting an uncertainty analysis on the leak detection methodology developed on the University of Utah campus. This analysis will focus on quantifying uncertainty associated with variable monitoring scenarios such as the number of detection towers, location of detection towers, and various site characteristics associated with the monitoring effort. Furthermore, the data collected during the controlled release experiments described in the above paragraphs will be used to determine whether or not an inverse model can be built and used to effectively locate and quantify leaks based on data collected at the detection tower and machine-learning algorithms.

470601 A Numerical Simulation Update of the Aquistore CO_2 Storage Project

Abstract:

Aquistore serves as the storage site for the world's first commercial postcombustion carbon capture, utilization, and storage (CCUS) project from a coal-fired power generation facility. The operation site is located near SaskPower's Boundary Dam Power Station, just outside of Estevan, Saskatchewan, Canada. CO_2 injection and monitoring began in
April 2015.The Energy & Environmental Research Center (EERC), through its Plains CO_2 Reduction (PCOR) Partnership, is collaborating with the Petroleum Technology Research Centre (PTRC) and SaskPower to continue PCOR Partnership efforts to numerically model and interpret the performance of the project's injection well and observation well. The results of performance interpretation, modeling and simulation will improve MVA (monitoring, verification, and accounting) strategies based on history matching and predictive simulation results.

The deep saline system targeted for the Aquistore project comprises the Deadwood and Black Island Formations, the deepest sedimentary units in the Williston Basin. This paper describes the performance and interpretation of the injection and observation wells since the start of CO_2 injection, utilizing the dynamic simulation model and GEM – a compositional reservoir simulator from Computer Modelling Group Ltd. As of the end of April 2016, approximately 65,000 tonnes of CO_2 has been injected into the reservoir. The injection rate and length of the operating periods have been variable, but injection rate was increased near the end of 2015 to a level of 350 to 550 tonnes a day.

The injection and observation wells have been closely monitored, and history matching was performed while reconciling rate, pressure, temperature, variations in injectivity, injection flow profiles, and pulsed-neutron logging profiles. The current data set is well replicated in the simulation model by applying skin factor

adjustments and a local permeability reduction near the injection wellbore, possibly caused by a previous water injection test. However, variations in bottomhole pressure response remain considerable, and injection rate has experienced sudden large changes. Near the end of year 2015, injection rate increased along with an improved injectivity index, from a previous index value of 0.02 to 0.06 tonnes/day per kPa before November, to 0.08 to 0.16 tonnes per kPa after November, an increase of several hundred percent. This improvement is associated with a lower downhole temperature operating range.

CO_2 breakthrough at the observation well has been observed using pulsed-neutron logging, while the injection well flow profile indicates that channeling may have occurred and that CO_2 passes the assumed damage zone that limited the injectivity during early operations. A recent geophysical imaging attempt to observe the CO_2 plume will also be used for performance confirmations and to update the model correspondingly in order to reduce the uncertainty, thus providing more reliable performance prediction. Additional reporting of project performance may be appropriate in the future.

ACKNOWLEDGMENTS

This material is based on work supported by the U.S. Department of Energy National Energy Technology Laboratory under Award No. DE-FC26-05NT42592.

471125 Lessons Learned in Near-Surface Monitoring for Large-Scale CO_2 Storage

Abstract:

A 5-year near-surface monitoring program was developed and conducted by the Energy & Environmental Research Center as part of the Plains CO_2 Reduction Partnershipâ€™s large-scale (>1,000,000 tons CO_2/yr injection) CO_2 storage demonstration site. The purpose of the program was to establish baseline (preinjection) conditions and variability of surface water, soil, and shallow groundwater aquifer chemistries in the vicinity of geologic CO_2 injection that can be used in conjunction with continued assurance monitoring to 1) provide a scientifically defensible source of data to show that near-surface environments remain unaffected by fluid or gas migration and 2) identify and evaluate anomalies that could be indicative of an out-of-zone migration event should they occur during assurance monitoring.

The near-surface monitoring program implemented consists primarily of acquisition and analysis of soil gas and water chemistry data. This program comprised three parts: sampling of surface water features, sampling of shallow groundwater aquifers, and sampling of soil gas in the shallow vadose zone. Assessment of the analysis data indicated that a high degree of natural variability within these systems is possible, especially at select sampling locations. However, the techniques proved to generate sufficient data to both detect and attribute anomalies to natural processes within these environments.

The lessons learned from execution of the program encompass proper establishment of baseline site conditions, effective and efficient assurance monitoring following the start of CO_2 injection operations, and critical data management of monitoring results. Baseline activities included site characterization and reconnaissance, establishment of sampling locations and techniques, and evaluation of results to determine natural variation. Assurance monitoring allowed for protocols to be developed to address any potential anomaly detection and for prioritizing near-surface

sampling frequency, types of samples and analyses conducted, and client/landowner relations. Data management required strict organization in preserving raw data, processing data during evaluation, and database development for housing and accessing results. Extensive quality assurance/quality control practices were also developed and integrated throughout all aspects of the monitoring program. The methodology developed and lessons learned will allow future carbon capture, utilization, and storage operators to make informed decisions regarding site-specific monitoring programs at other commercial-scale injection sites throughout the region.

465072 Economic Modeling of Carbon Dioxide for Geologic Sequestration and Enhanced Oil Recovery in Jacksonburg-Stringtown Oil Field, West Virginia, USA

Abstract:

Continued global climate change caused by the use of fossil fuels may require effective control or mitigation of greenhouse gases. Geological carbon dioxide (CO_2) capture, utilization and storage (CCUS) is one of the three options to limit CO_2 emissions into the atmosphere. Injection of CO_2 into depleted oil and gas reservoir coupled with enhanced oil/gas recovery (EOR/EGR) is an economic-feasible and short-term methodology. CCUS costs encompass the full chain of CCUS process (i.e., capture, transport, storage and monitoring, including any utilization of CO_2 that results in its long-term sequestration). In order to harmonize the costs of CCUS, a common cost-benefit model is developed to evaluate the economic feasibility of geologic CO_2 storage implemented in conjunction with enhanced oil recovery (EOR) in Jacksonburg-Stringtown oil field in West Virginia. The cost-benefit analysis framework developed in this study includes a reservoir performance model integrated with a cost model and a discounted cash flow economic model. Realization of the economic incentives is associated with CO_2-EOR storage operations in Jacksonburg-Stringtown oil field. This model will improve the clarity and consistency of cost estimation for CCUS and may prove the impetus for commercial scale deployment of CCUS systems from the numerous coal-fired electric power plants in the Ohio River region of the Appalachian basin.

455372 Carbon Dioxide-Induced Liberation of Methane from Laboratory-Formed Methane Hydrates: A Pathway to Economical Carbon Sequestration

Abstract:

The presence of vast methane (CH_4) hydrate reserves around the globe are known for decades. Gas production by external stimulation techniques such as steam injection, gas depressurization or hydrate-inhibitor injection is well-established but an economical production method remains a challenge. In parallel, there is a sense of urgency to minimize carbon dioxide (CO_2) levels in the atmosphere that have exceeded 400 ppm. Both CH_4 and CO_2 gases form structure sI hydrates and at temperatures below 10°C, CO_2 hydrates are more thermodynamically stable at lower pressures than CH_4 hydrates. Considering that burying CO_2 in deep formations is one of the carbon sequestration options being considered, the CH_4-CO_2 exchange in natural methane hydrate reservoirs may be a way to address both issues. The CH_4-CO_2 exchange option was attractive enough to warrant a commercial test by ConocoPhilips in 2012 in which over 24,210 m^3 of CH_4 was released when a 23 mol% CO_2 in N_2 mixture was injected in a hydrate reservoir in Alaska. Carbon dioxide was preferred over N_2 in the hydrate phase, as about 70% of the N_2 gas injected was recovered, while only 40% of the 1376 m^3 of CO_2 injected was retrieved. Thus, the CH_4-CO_2 exchange reaction merits further research.

In this talk, we will describe several laboratory-scale experiments to understand the CH_4-CO_2 hydrate exchange kinetics. Five hydrate formation-decomposition runs focused on CH_4-CO_2 exchange, two baselines and three with host sediments, were performed in a 200 mL high-pressure Jerguson cell fitted with two glass windows that allowed visualization of the time-resolved hydrate phenomenon. The data show that the induction time for hydrate appearance was the largest at 96 h with CH_4 while with CO_2, the time shortened by a factor of four. However, when the secondary gas (CO_2 or CH_4) was injected into the system containing preformed hydrates, the entering gas formed the hydrate phase instantly (within minutes) and no lag was observed. In a system containing host Ottawa sand (104 g) and artificial seawater (38 mL), the induction period reduced to 24 h. The CO_2 hydrate formation in a system that already contained CH_4 hydrates was facile and they remained stable, whereas CH_4 hydrate formation in a system consisting of CO_2 hydrates as hosts were initially stable, but CH_4 gas in hydrates quickly exchanged with free CO_2 gas to form more stable CO_2 hydrates. In all five runs, even though the system was depressurized, left for over a week at room temperature, and flushed with nitrogen gas in between runs, hydrates formed instantaneously, irrespective of the gas used. The â€œmemory effectâ€ was exhibited with either gas, a result in contradiction with that reported previously in the literature. The facile CH_4-CO_2 exchange observed under temperature-pressure conditions that mimic naturally-occurring CH_4 hydrates show promise to develop a commercial carbon sequestration system.

458610 Techno-Economic Assessment of CO_2 Quality Effect on Its Storage and Transport (CO2QUEST)

Abstract:

The ultimate composition of the CO_2 stream captured from fossil fuel power plants or other CO_2intensive industries and transported to a storage site using high pressure pipelines will be governed by safety, environmental and economic considerations.

So far, most of the studies performed on this topic have been limited in scope, primarily focusing on investigating the impact of the CO_2stream impurities on each part of the Carbon Capture and Sequestration (CCS) chain in isolation. This is a significant drawback given the markedly different sensitivities of the pipeline, well bore materials and storage sites to the various impurities.

For example, given the risk of water table contamination, trace elements such as Lead, Mercury and Arsenic in the CO_2 stream are of far greater concern in an aquifer storage site than compared to the pipeline. On the other hand, even small concentrations of water in the CO_2 stream are detrimental to the pipeline due to corrosion, but of benefit even at high concentrations during storage given the immobilisation effect of water on CO_2.

'*What is good for the pipeline is not necessarily good for storage'.*

It is clear that the optimum composition and concentration of the impurities in the captured CO_2 stream involves a delicate balance between the different requirements within the CCS chain, spanning capture, transportation and storage, with cost and safety implications being the over-arching factor. Pivotal to these considerations is an understanding of the impact of the impurities on the physico-chemical properties of CO and its hazard profile.

This paper presents and overview and the main findings of the FP7 European Commission CO2QUEST project. Involving the collaboration of 12 industry and academic partners in Europe, China and Canada the project addresses the fundamentally important and urgent issues regarding the impact of the typical CO_2streams impurities captured from fossil fuel power plants on its safe and economic transportation and storage.

The work programme, spanning 36 months, focuses on the development of state-of-the-art mathematical models, backed, by laboratory and industrial-scale experimentation using unique EC-funded test facilities to perform a comprehensive techno-economic, risk-based assessment of the impact of the CO_2stream impurities on the phase behaviour and the physico-chemical reactions governing the pipeline and storage site integrities.

The above involves the determination of the important CO_2 mixtures that have the most profound impact upon the pipeline pressure drop, compressor power requirements, pipeline propensity to ductile and brittle facture propagation, corrosion of the pipeline and well bore materials, geochemical interactions within the well bore and storage site, and the ensuing health and environmental hazards. Based on a cost/benefit analysis and whole system approach, the results are employed to provide recommendations for tolerance levels, mixing protocols and control measures for pipeline networks and storage infrastructure thus contributing to the development of relevant standards for the safe design and operation of CCS.

461978 Understanding the Factors That Affect CO_2 Transport, Storage, and Enhanced Oil Recovery Potential in Unconventional Reservoirs

Abstract:

Investigations on the use of subsurface geologic formations for carbon dioxide (CO_2) storage have been ongoing for many years and have traditionally focused on saline aquifers, depleted oil and gas formations, and unminable coal seams (Goodman and others, 2014; U.S. Department of Energy National Energy Technology Laboratory, 2012). In the last decade, improvements in directional drilling technology (Watney, 2016) and hydraulic fracturing (King, 2010) of organic-rich shale and other tight rock formations have opened new opportunities for CO_2 storage, with the potential cobenefit of enhancing gas and oil recovery in reservoirs that typically have less than 10% primary recovery.

Past and ongoing work at the Energy & Environmental Research Center (EERC), funded by the U.S. Department of Energy National Energy Technology Laboratory in conjunction with partners from the oil and gas industry, has been focused on better understanding the factors that affect CO_2 transport as it relates to both enhanced oil recovery (EOR) and CO_2 storage in unconventional oil reservoirs and organic-rich shales. Much of this work has been performed in several of the EERC's laboratories using rock samples from the Bakken petroleum system, the results of which can be applied to unconventional formations across the United States. The analytical work conducted to date has focused on supercritical CO_2 extraction experiments, CO_2 sorption experiments, and extensive physical and geochemical characterization using rock samples from several Bakken wells to better understand the factors that can affect CO_2 migration, oil extraction, sorption, and storage. Sample characterization has been accomplished using a variety of analytical techniques including whole-core and microcomputerized tomography imaging, conventional scanning electron microscopy (SEM) coupled with energy-dispersive x-ray spectroscopy, field emission SEM, focused ion beam SEM, x-ray fluorescence, x-ray diffraction, optical microscopy, Rock-Eval pyrolysis, vitrinite reflectance, mercury injection capillary pressure testing, and geomechanical testing. Initial characterization work focused on better understanding the occurrence and characteristics of fracture and pore networks and has expanded to the use of advanced analytical and image analysis techniques to quantify fracture and pore occurrence and mineralogical content at various resolutions. Work is ongoing to better understand and quantify the occurrence and accessibility of pores, fractures, clay minerals, and kerogen and the potential implications that those features have on CO_2 transport for EOR and on CO_2 sorption for storage. The sample characterization efforts are being used to better understand the results of supercritical CO_2 extraction experiments and ongoing CO_2 sorption experiments using Bakken samples that have undergone extensive characterization. These results are being integrated into modeling research activities to determine the fundamental mechanisms controlling fluid transport and CO_2 storage in the Bakken, resulting in the development of improved model simulations. The results of the EERC's laboratory-based work and its implications for CO_2 EOR and storage, as well as future research needs, will be discussed.

REFERENCES

Goodman, A., Fukai, I., Dilmore, R., Frailey, S., Bromhal, G., Soeder, D., Gorecki, C., Peck, W., Rodosta, T., and Guthrie, G., 2014, Methodology for assessing CO_2 storage potential of organic-rich shale formations: Energy Procedia, v. 63, p. 5178–5184.

King, G.E., 2010, Thirty years of gas shale fracturing: what have we learned? Society of Petroleum Engineers: Annual Technical Conference and Exhibition, September 19–22, Florence, Italy, SPE 133456.

U.S. Department of Energy National Energy Technology Laboratory, 2012, The United States 2012 Carbon Utilization and Storage Atlas, Fourth Edition: www.netl.doe.gov/File%20Library/Research/Coal/carbon-storage/atlasiv/Atlas-IV-2012.pdf (accessed January 2016).

Watney, W.L., 2016, Horizontal drilling – technology review, current applications, and its future in developing Kansas' petroleum resources: Kansas Geological Survey, Lawrence, Kansas, www.kgs.ku.edu/PRS/Unger/Watney-KBA-KIOGA--Horizontal_Drilling.pdf (accessed January 2016).

470324 Evaluating the Feasibility and Scale of Geologic CO_2 Storage Potential in Ohio's Appalachian Region

Abstract:

Evaluating the feasibility and scale of geologic CO_2 storage potential in Ohio's Appalachian Region

Priya Ravi Ganesh*, Autumn Haagsma, Glenn Larsen, Isis Fukai, Joel Main, Mackenzie Scharenberg, Neeraj Gupta

Battelle, Columbus, OH

Around the world, various industrial-scale carbon dioxide (CO_2) geologic storage (GS) demonstration projects have been or are being conducted in deep saline formations. To fully tap into the overall potential of CO_2 GS resources, comprehensive research focusing on the detailed geologic characterization of potential regions, such as the Appalachian Basin in the Midwestern United States, needs to be completed. The project described here is part of a long-term program to investigate CO_2 GS potential in Ohio and adjacent areas. Here, we apply an integrated multi-scale approach to evaluate the possible implications of basin-scale pressure buildup on storage capacity estimation. The reservoir analysis task is divided into two main efforts: (1) *Static capacity assessment* to help provide regional estimates as well as specific site estimates for selected saline storage site(s) and (2) *Dynamic capacity and injectivity assessments* of selected saline storage site(s) as illustrative examples for the evaluation of injectivity and capacity from the study area. The static and dynamic estimates for CO_2 storage shall be compared to highlight the effects of dynamic operational constraints such as rates, well locations, etc. typical of realistic CO_2GS site conditions.

Our study region consists of ~61,000 sq.km of complex geology in eastern Ohio, with a primary emphasis on the strata that underlie the Lower Ordovician Knox unconformity erosional surface. These strata are the Beekmantown dolomite through the Mt. Simon/basal sandstone where many of these deeper porous layers contain highly saline brines that have remained isolated from shallower zones. Key inputs to this feasibility study draw from the geologic characterization and fracture gradient assessment activities in the study area done as part of other tasks in this project. The static resource capacity estimation shall be determined with considerations for the total storage efficiency of CO_2 in these modeled saline formations. The potential storage sites determined from the geologic characterization exercise shall be evaluated to understand dynamic CO_2 GS performance metrics of interest, namely, injectivity, pressure buildup and spatial distribution of injected CO_2volume through time.

This study aims to help evaluate the feasibility and scale of CO_2 GS in regions in Ohio with respect to performance metrics of storage capacity and CO_2 injectivity. We work to evaluate different scales such as the regional or formation scale, plume or site-scale, as well as local-scale or near-well effects of CO_2 injection. This multi-scale study considers different resolution of input detail such as porosity and permeability distributions and uses key performance metrics, such distribution of injected CO_2 and pressure, to provide insights into acceptable site spacing, pressure buildup and CO_2plume distribution.

At the formation or regional level, we begin by mapping static storage resource estimates for each formation to create a roadmap to CO_2 GS in eastern Ohio. CO_2 GS resource is defined as the mass estimate of CO_2 that can be stored in a given geologic formation. We use different levels of data resolution for input formation properties (e.g., porosity and thickness) for this exercise. These range from the knowledge of average formation properties to the availability of the spatial distribution of these properties. Results show that, in descending order, the Mt. Simon/ basal sandstone, Kerbel sandstone, and Rose Run sandstone have the highest specific CO_2GS resource capacities. We investigate the subsurface dynamics by considering multiple simultaneous operational storage sites scenarios in selected study areas with high storage potential for these formations. The dynamic numerical modeling scenarios that consider multiple operational sites in a given formation or region shall provide in insights into acceptable site spacing from the extent of the pressure front to minimize pressure interference between sites.

The study areas for CO_2 GS in Cambrian formations in the eastern Ohio study area were determined from the formation maps created for the study area. Locations where the petrophysical indicators for storativity (porosity-feet) and transmissivity (permeability-feet) highs overlap are considered for further detailed study for CO_2 storage with dynamic assessment. The top three sandstone units identified from the static storage resource estimation exercise have been analyzed to identify 'sweet spots'. At the plume or local-scale examination for a given formation, these individual sites shall be modeled with possible CO_2 injection well patterns simulated. The results from this exercise test the effect of pressure buildup and potential interference issues (well spacing), as well as tracking the injected CO_2plume extent.

On a finer scale, a single well at the center of the possible well patterns for each site in a given formation are simulated for 30 years of CO_2injection, followed by 50 years of post-injection monitoring, to get insights into local-scale near-wellbore dynamics. These simplified well models with equivalent average site properties in radial configurations are being tested to help with preliminary sensitivity analysis to relate to monitoring aspects during injection operations. This exercise establishes sensitivity to well configurations, relative permeability etc. representative of local-site conditions in the chosen detailed study site(s).

The integration of results from these different scales are expected to bring together insights into the practical feasibility of commercial CO_2 GS storage in our study region. Our study aims to provide insight into the CO_2 GS potential for deep saline formations in eastern Ohio while reflecting the effects of dynamic operational constraints typical of realistic CO_2GS site conditions in the study region.

This work was supported by the Ohio Development Services Agency OCDO Grant OOE-CDO-D-13-22 and the U.S. Department of Energy through the Midwest Regional Carbon Sequestration Partnership award number DE-FC26-05NT42589. In-kind contributions and well access was provided by a number of well operators in the region.

460550 Shallow Ground Water Monitoring at the Secarb CO_2 Injection Site Near Citronelle, Alabama

Abstract:

The SECARB Anthropogenic Test is a fully integrated U.S. Department of Energy sponsored-research project located near Citronelle Alabama designed to demonstrate the safe, reliable capture, transportation and storage of CO_2. Starting in August 2012, the project team injected over 114,000 metric tonnes of CO_2 into the Paluxy Formation at a depth exceeding 2,865 m over a two-year period. Injection concluded in September 2014 and the project is currently in the post-injection site care period. Several monitoring techniques are being used to demonstrate post-injection containment of the CO_2 in the injection zone and non-endangerment of groundwater resources during the three-year post-injection monitoring effort. These techniques include time-lapse cross-well and vertical seismic profile surveys, geophysical logging, reservoir pressure and temperature measurements, soil CO_2 flux measurements, tracer surveys, and shallow groundwater sampling and analyses.

The ground water monitoring program consists of collecting water samples from three dedicated monitoring wells and one water supply well installed in the shallow surficial aquifers in close proximity to the projectâ€™s deep CO_2 injection and observation wells. A risk assessment performed during the early stages of the project identified the deep wellbores, which penetrate the confining unit overlying the injection zone, as the only potential migration pathway. No known faults are present based on available geologic and seismic data. Although the risk of CO_2 leakage is expected to be extremely small, SECARB installed the shallow ground water monitoring wells on the injection well pads to check for a CO_2 release as part of the regulatory permitting program.

Water samples are collected from each monitoring well on a quarterly schedule. Samples are analyzed on-site for time-sensitive parameters including pH, electrical conductivity, dissolved oxygen, temperature and oxidation-reduction potential and preserved for off-site analysis for 17 metals, total inorganic carbon (TIC) and other inorganic parameters by a contract laboratory. The ground water analytical results are compared to U.S. Environmental Protection Agency (EPA) drinking water standards and to background samples collected prior to CO_2 injection using EPA-approved statistical methods.

The study results indicate that shallow ground water systems are dynamic and complex, resulting in water quality results for individual naturally-occurring constituents that can fluctuate considerably with time due to natural processes or sampling-induced mobilization of inorganics. This may lead to false detection and reporting of a CO_2 release over a 40 to 100-year life span of a commercial CO_2 storage project. Therefore, multiple constituents (e.g., pH, Fe, TIC and EC), or lines of evidence, should be collectively used in the analysis to detect a release and not just one or two (e.g., pH). The duration and location of background sampling should also be carefully considered in the initial design of the storage facility due to the ubiquitous occurrence of inorganics found naturally in ground water.

471378 Probabilistic Analysis of CO_2 Storage Mechanisms in a CO_2-EOR Field Using Polynomial Chaos Expansion

Abstract:

Depleted oil fields are already used for storing CO_2 via injection of CO_2 for enhanced oil recovery (CO_2-EOR). Such storage is an outcome of CO_2-EOR, albeit not necessarily by design. A next step would be intentional storage via post-EOR CO_2 injection. Following injection for EOR or post-EOR, CO_2 may be trapped by dissolution in the oil or aqueous phases, or remain in the supercritical CO_2 phase. Forecasting the nature of trapping and the ultimate CO_2 distribution in a reservoir is hindered by uncertainty in reservoir properties. The purpose of this study is to develop and apply reduced order models (ROMs) integrated with Monte Carlo simulations to quantify the three primary types of storage mechanisms in an active CO_2-EOR field, including oil solubility trapping (oil phase), aqueous solubility trapping (aqueous phase), and hydrodynamic trapping (CO_2 in supercritical phase). The case study site for this analysis is the SACROC unit in western Texas. A Polynomial Chaos Expansion (PCE) technique was used to develop the ROMs. The sources of uncertainty considered are porosity and permeability. Model outputs of interest include dissolved mass of CO_2 in oil phase, mass of CO_2 in supercritical phase, dissolved mass of CO_2 in aqueous phase, and oil saturation in the rock formation. Reduced order models were developed for all cells in five selected layers of the model, which are adjacent to injection wells, at three specific time points of interest, including the end of a simulated CO_2-EOR period, the end of a post-EOR continuous CO_2 injection period, and the end of a post-injection monitoring period. Results of regression fit and validation analysis yield high R^2 values and low NRMSE values, indicating that the ROMs derived from PCE are able to make meaningful predictions (compared to conventional reservoir models) and effectively represent relationships between model inputs and model outputs of interest. Results of 1000 Monte Carlo simulations indicate a dominant upward transport of CO_2 during injection, driven by buoyancy, and a dominant

downward transport of dissolved CO_2 after injection stops, caused by the increase in both oil and brine density. At the end of the 100-year simulation of the SACROC case study, the selected five layers store between 54% and 61% of total trapped CO_2 in the entire domain in the oil phase, between 21% and 24% in the supercritical phase, and between 11% and 15% in the aqueous phase. The selected five layers store between 89% and 98% of total trapped CO_2 in the entire domain by the end of simulation.

465751 Monitoring Ions Based on Reactive Transport Modeling of CO2 Sequestration in Ulleung Basin, Korea

Abstract:

In Carbon Capture and Storage(CCS) process geochemical monitoring technology plays important role to avoid potential CO2 leakage. In this paper, reactive transport modeling of supercritical CO2 in aquifer which is modeled after Ulleung basin; one of offshore geological storage site of Korea was performed. Supercritical CO2 have been injected into target reservoir at the rate of 30kg/s for 30 years. After injection for 30 years, long-term behavior of the CO2 by considering its geochemical reaction for 5000 years was investigated. In early stage, aqueous phase and supercritical CO2 was dominant. After 800~1000years, mineral phase became dominant phase accounts for more than 80% of injected CO2. During injection period, at monitoring points, manganese, bicarbonate ions concentration changed before reaching supercritical CO2. Ion concentration changes can give early warning of potential CO2 leakage. To determine monitoring ions for specific mineral composition, reactive transport modeling of typical mineral composition cases for geological storage reservoir all over the world was conducted.

459744 Multiple Modes of Emplacement of Rare Earth Elements in the Fire Clay Coal, Eastern Kentucky:Â Implications for Utilization

Abstract:

Rare earth elements and yttrium (REY) in coal deposits have attracted much attention in recent years because of their potential economic significance. Based on previous studies (e.g., Seredin and Dai, 2012), four possible modes for emplacement of the REY are considered: terrigenous, tuffaceous, infiltrational, and hydrothermal, with eastern Kentuckyâ€™s Pennsylvanian-age Fire Clay coal considered to be a typical example of the tuffaceous mode. The Fire Clay coal owes much of its high REY content to REY-bearing zircon and phosphates in a volcanic-ash-fall tonstein within the coal in many locations. However, some of the original REY elements may have been terrigenous detrital minerals deposited in the peat. Leaching of REY from the tonstein into the surrounding coal and the hydrothermal overprint of mineralizing fluids associated with the northwestwardly movement of the Pine Mountain thrust sheet contributed to the total REY signature in the Fire Clay coal. Each type of REY emplacement has the potential to create different patterns of elemental distribution. Not all coals are going to have the complex history of the Fire Clay coal, but it should be considered that the total history of REY emplacement by multiple mechanisms contributes to both the total REY concentration and the relative distribution of the individual lanthanide elements. In the beneficiation of the mined raw coal feed, certain portions of the mined coal will be removed from the clean product, meaning that the product has a different whole-coal REY total and REY distribution than the mined material.

466773 Process Design and Economics of a Novel Process for the Extraction of Rare Earth Elements from Coal Ash

Abstract:

Rare earth elements are used extensively in military hardware, high-end aircraft equipment, refinery catalysts and the clean energy industry. However, the rare earth mineral resources in the U.S. are limited, with 95% of their supply coming from foreign imports.[1] In this context, coal ash is an abundant waste material from coal power plants with high content in rare earth elements (REE), particularly in heavier REE (e.g., yttrium, dysprosium, and others) important in high performance magnets, flat panel displays, compact fluorescent lighting, LEDs, and lasers. Coal ash can be a promising, low-cost resource of heavy and light rare earths.[2] Moreover, coal ash contains other valuable components such as carbon and magnetite, and is widely utilized as cement substitute in concrete and construction materials.[3] Therefore, it is useful but also challenging to design a chemical plant to process coal ash for recovering the REE, as well as other valuable byproducts.

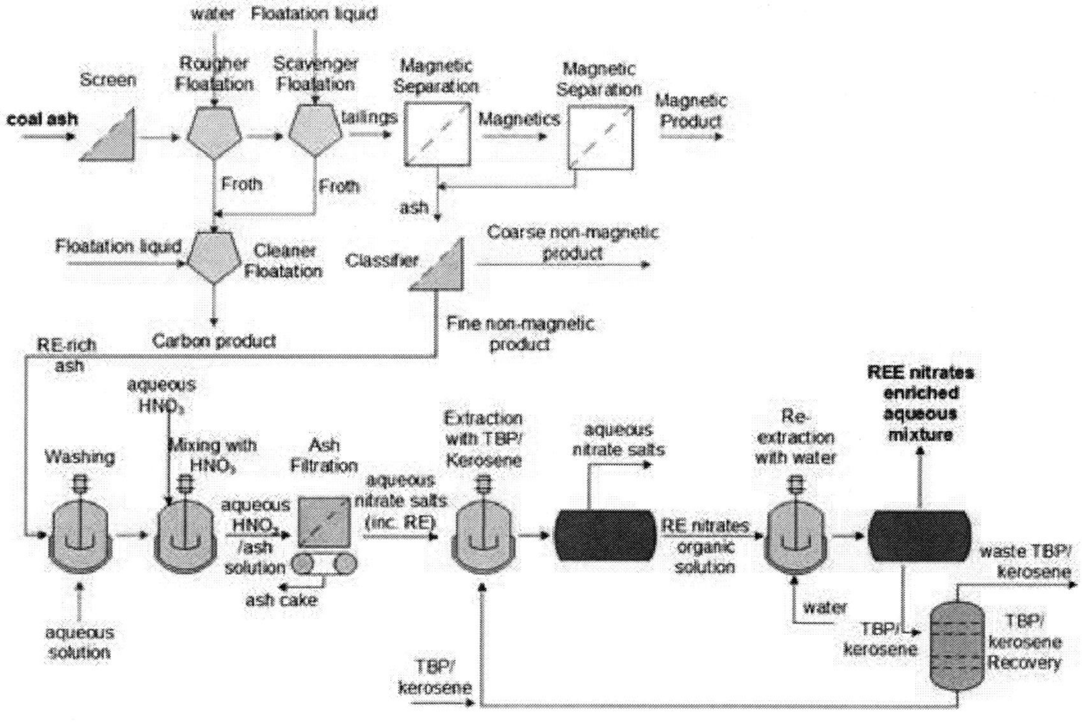

Figure 1: Simplified diagram of physical and chemical enrichment of REE from coal ash.

Our process for REE production from coal ash involves two main stages: a physical enrichment step and a chemical enrichment/extraction step. The process comprises several key steps, shown in Figure 1, including screening, flotation, magnetic separation, size separation, aqueous washing, nitric acid digestion, and solvent extraction. Batch laboratory-scale tests of each of these steps, as well as continuous flow test for key steps, were first carried out. Then, a steady state model of a continuous ash processing plant was developed in ASPEN Plus. The model was used for the simulation of the production capacity and efficiency of REE extraction, as well as the estimation of the streams of coproducts produced, such as carbon, magnetite, and the ashcake as cement substitute. The model was first matched with the laboratory results and then scaled up to a high throughput plant. Subsequently, economic analysis was performed to examine the –commercial feasibility of the process and identify the most cost-sensitive and challenging steps that need to be researched further. The analysis showed that the process is profitable with sufficient recycling of the acid and organic solutions. Several case studies were explored for the process economics, by varying the price of the REE concentrate and the cost of recycling the reagents to assess the product/coproduct pricing and profitability space.

Acknowledgment

This work was supported by the Office of the Secretary of Defense (OSD) with Dr. David Shifler of the Office of Naval Research (ONR) as the Program Manager under Contract N00014-15-C-0039.

References

[1] Critical Materials Strategy, U.S. Department of Energy, December 2010 and December 2011.

[2] Seredin, V. V.; Dai, S. Coal deposits as potential alternative sources for lanthanides and yttrium. *International Journal of Coal Geology*, 2012, *94*, 67.

[3] Blissett, R.S.; Rowson, N.A. A review of the multi-component utilisation of coal fly ash. *Fuel*, 2012, *97*, 1.

474522 Characterization and Recovery of Rare Earths from Coal and By-Products

Abstract:

Coal is a precious resource, both in the United States and around the world. The United States has a 250-year supply of coal, and generates between 30 - 40% of its electricity through coal combustion. Approximately 1 Gt of coal has been mined annually in the US, although the 2015 total will likely be closer to 900 Mt (http://www.eia.gov/coal/production/quarterly/). Most of the coal is burned for power generation, but substantial quantities are also employed in the manufacture of steel, chemicals, and activated carbons. Coal has a positive impact upon many industries, including mining, power, rail transportation, manufacturing, chemical, steel, activated carbon, and fuels.

Everything that is in the earthâ€™s crust is also present within coal to some extent, and the challenge is always to utilize abundant domestic coal in clean and environmentally friendly manners. In the case of the rare earths, these valuable and extraordinarily useful elements are present within the abundant coal and coal by-products produced domestically and world-wide. These materials include the coals, as well as the combustion by-products such as ashes, coal preparation wastes, gasification slags, and mining by-products. All of these materials can be viewed as potential sources of rare earth elements.

Most of the common inorganic lanthanide compounds, such as the phosphates found in coal, have very high melting, boiling, and thermal decomposition temperatures, allowing them to concentrate in combustion and gasification by-products. Furthermore, rare earths have been found in elevated concentrations in the strata above and below certain coal seams.

The National Energy Technology Laboratory (NETL) recently initiated research for the determination and recovery of rare earths from abundant domestic coal by-products. The NETL Rare Earth EDX Database (https://edx.netl.doe.gov/ree/) is a resource for rare earth information as related to coal and by-products. Many other research organizations have also initiated efforts for the determination and recovery of rare earths from unconventional sources such as coal by-products.

Much of the recent research on coal utilization in the United States has focused upon the capture of pollutants such as acid gases, particulates, and mercury, and the greenhouse gas carbon dioxide. The possible recovery of rare earth and other critical elements from abundant coal and by-products is an exciting new research area, representing a dramatic paradigm shift for coal. Additional data is needed on the rare earth contents of coals and by-products in order to determine the most promising potential feed materials for extraction processes. Future work will focus on characterization of coals and by-products, and separation/partitioning methods for rare earth recovery. Some of the early and exciting NETL-RIC results for characterization, recovery, and the geochemistry of the rare earths and other critical elements will be briefly described.

474523 Reactive Grinding of Coal Ash for Enhanced Rare Earth Extraction

Abstract:

The size of rare earth minerals in coal and coal by-products tend to be on the order of 1 - 20Î¼m, and intensive grinding may be required to liberate these minerals for subsequent beneficiation and extraction. By adding reactants to the grinder, such as sodium hydroxide, the rare earth minerals could react to form rare earth hydroxides, which are more amenable to leaching in acid. Coal ash was subjected to grinding in a standard laboratory ball mill with, and without reactants, such as sodium hydroxide, in order to enhance subsequent leaching of the rare earth elements in hydrochloric acid.
In the case of sodium hydroxide, it is hypothesized that the rare earths in the coal ash are converted to rare earth hydroxides, which are more amenable to leaching in hydrochloric acid verses rare earth phosphates. Various grinding times were studied to determine the extraction yield of the rare earth elements. Grinding materials to increase overall surface area of the sample can also increase the leaching efficiency, and this was also examined. Grinding fly ash with 25wt% NaOH for 3 hours showed significantly enhanced extraction of the rare earth elements compared to grinding for 3 hours without NaOH. The extraction of other elements such as phosphorous will also be discussed.

474524 Distributions and Associations of Rare Earth Elements in Fly Ash Using Laser Ablation High Resolution Mass Spectrometry

Abstract:

Laser ablation high resolution mass spectrometry (LA-HR-MS) was used to study the distributions and associations of rare earth elements with other major and minor elements in coal fly ash samples using line scans and statistical analysis. Associations of the REEs with major elements, which compose different components of a fly ash, were analyzed. The results indicate that there are good positive correlations amongst the light rare elements (LREE) and thorium, which is a possible indication of monazite, and a good correlation between the heavy rare earth elements (HREE) and yttrium, which is an indication of xenotime. Associations of the rare earths with elements in the glass fraction and volatile elements usually found on the surface of fly ash particles will be discussed. Dissimilar correlations of elements with the LREE and HREE may indicate differences in the thermal stabilities of source REE minerals, and may also help determine the location and best extraction method for recovery of LREEs and HREEs in coal and coal ash.

455274 Frother Performance and Its Influence on Flotation Process

Abstract:

2016 AIChE Annual Meeting
November 13th - November 18th, 2016
Hilton San Francisco Union Square
California, CA

Abstract Submission

PROGRAM TOPIC:

Rare Earth Elements in Fossil Fuel Derived Solids and Liquids

Exact Title of Paper:

Frother performance and its influence on flotation process

Presenting Author: Yungchieh Lai

National Energy Technology Laboratory, Department of Energy, Pittsburgh, PA

Contact Information: Yungchieh.lai@netl.doe.gov

Co-Authors: Fan Shi

National Energy Technology Laboratory, Department of Energy, Pittsburgh, PA

Contact Information: Fan.Shi@netl.doe.gov

Co-Authors: Yee Soong

National Energy Technology Laboratory, Department of Energy, Pittsburgh, PA

Contact Information: Yee.Soong@netl.doe.gov

Co-Authors: McMahan Gray

National Energy Technology Laboratory, Department of Energy, Pittsburgh, PA

Contact Information: Mac.Gray@netl.doe.gov

Abstract:

Flotation process is an aqueous, efficient, inexpensive, and well-developed commercial technology and has been widely applied in a variety of industrial processes, including mineral processing and waste water treatments. The flotation process first generates gas bubbles within a solid-liquid or liquid-liquid suspensions. The gas bubbles then attach solid particles or immiscible liquid droplets and rise to top of the liquid surface. The floated solid particles or immiscible liquid droplets finally are skimmed off to achieve the separation purpose. As particles/immiscible droplets are hydrophobic, such as greases and oils, they are easily attachable to gas bubbles and can be separated by the flotation technology.

A successful flotation process is dependent on the probability of the collision between bubbles and particles or droplets. Gas holdup is the ratio of gas volume to liquid phase in a bubble device. The increase of gas holdup will increase the opportunities of contact and attachment between particles and gas bubbles and hence increase the flotation efficiency.

The mechanic agitated flotation cell is one of the major flotation devises available, and frothers is the key to generate high gas holdup in such devise. Currently, the flotation technology has been applied at very different conditions, such as ion-containing oily waste water for oil cleanup or low temperature area for mineral processing. The effectiveness of frothers therefore can be strongly influenced by these different flotation conditions.

Here, we report the performance of frothers in the mechanic agitated flotation cell. Four different types of frothers- commercial product F660, commercial product F672, 4-Methyl-2-pentanol, and 4-heptanol- were studied at different conditions for their gas holdup generation. We will discuss the changes of gas holdup with the changing conditions. Flotation tests for practical separations are planned for the coming future. The discussion for the correlation between the performance of frothers and the separation efficacy at different conditions will be included.

451108 Enrichment of Rare Earth Elements from Coal and Coal By-Products By Physical Separations

Abstract:

Rare earth elements (REEs) are important elements used in a variety of high-technology products and thus play a critical role in economic growth and national security. To diversify the supply chain of these critical materials, the U.S. Department of Energy National Energy Technology Laboratory is investigating technologies and processes for the recovery of REEs from America's vast coal resources. In this work, size, density and electromagnetic separations of various coal and coal by-product samples were carried out to investigate the feasibility of using these physical separation methods for REE enrichment. The samples include fly ash, bottom ash, clean coal, coal refuse and clay. Samples were analyzed using ICP-MS, ICP-OES, TGA and CHNS methods. Preliminary results indicate that the REE contents in fly ash generally increase as the ash particle size decrease and REEs are enriched in the non-magnetic fractions. The results also suggest that density separation is the most effective method for REE enrichment compared with other methods tested in this work. The combination of these physical separation methods would be beneficial and are being further pursued in this research.

AUTHOR INDEX

AUTHOR INDEX

AUTHOR INDEX